EUROPEAN DOCKLANDS

Past, present and future

An illust
to glorio
splendid
dramatic
in Europ

Front Cover
*Antwerp's picturesque old waterfront
circa 1656 with its fine churches and
buildings.(Grootserminarie Brugge)*

Back Cover
*Copenhagen's beautiful waterfront
today with its rich architectural and
historical landmarks.(Port of
Copenhagen Authority)*

**Edited by S K Al Naib
Polytechnic of East London**
with the assistance of
international experts

ISBN 0-901 987-82-4

First Printing   March 1991

Other books by the same editor
"Dockland"
"London Docklands"
"Hydraulic Structures" (forthcoming)

This is a companion volume to
'London Docklands'

by the same editor

Typeset By ILEA (CRS) London

Printed by Ashmead Press Ltd., London

Preface

European Docklands is an illustrated account of history, heritage and urban regeneration in a number of old ports. The book traces the splendid stories of these dockyards over the past two centuries. Currently, these old docks and warehouses are being converted into office complexes, luxury homes and leisure centres.

By working together with an international team of researchers and many specialist organisations, we were able to produce this fascinating book. Topics include history, surviving warehouses, museums, docklands transport, present transformation and future development into the twenty-first century.

A major new book on historic docklands provides a vivid portrait of development from the eighteenth century to the present day. It contains over 70 illustrations, many of which provide a lasting memory for those who love these old places with their enchanting warehouses and vast water areas.

Today's docklands are undergoing enormous changes. Their heritage being a vital part of our history, provide a foundation on which the future of these areas can safely be built.

Transformations have taken place in most European Ports. The old docklands often located close to the city centre, are no longer used for port activities and offer a major opportunity for regeneration and economic revival. Utilizing the redundant dock water areas as unique environmental features, new commercial and residential projects are emerging along the old docks.

Originally, seven major North Sea channel ports were selected: London, Calais, Antwerp, Rotterdam, Hambourg, Copenhagen and Gothenburg. Following further consultation locally and the availability of contributors, Calais was replaced by Cherborg and Hambourg by Duisburg. London Docklands have been published in aseparate bookby the same editor, and the remaining six ports are covered in this companion volume.

This book provides an authoritative and unique reference for the general public, port authorities, city councils, private and commercial developers, Government departments and other organisations. It is suitable for use in schools and colleges to study and compare the history and development of representative European docklands and the way in which their futures will reflect the past of these historically important areas.

An invitation to the reader

If you are a reader with knowledge and unpublished information on docklands in any part of the world or with interesting stories on life and work in London docklands, or can assist in any Docklands research, please contact:

Dr S K Al Naib
Head of Department of Civil Engineering
Polytechnic of East London
Longbridge Road, Dagenham
Essex, RM8 2AS
England

Telephone: 081 590 7722

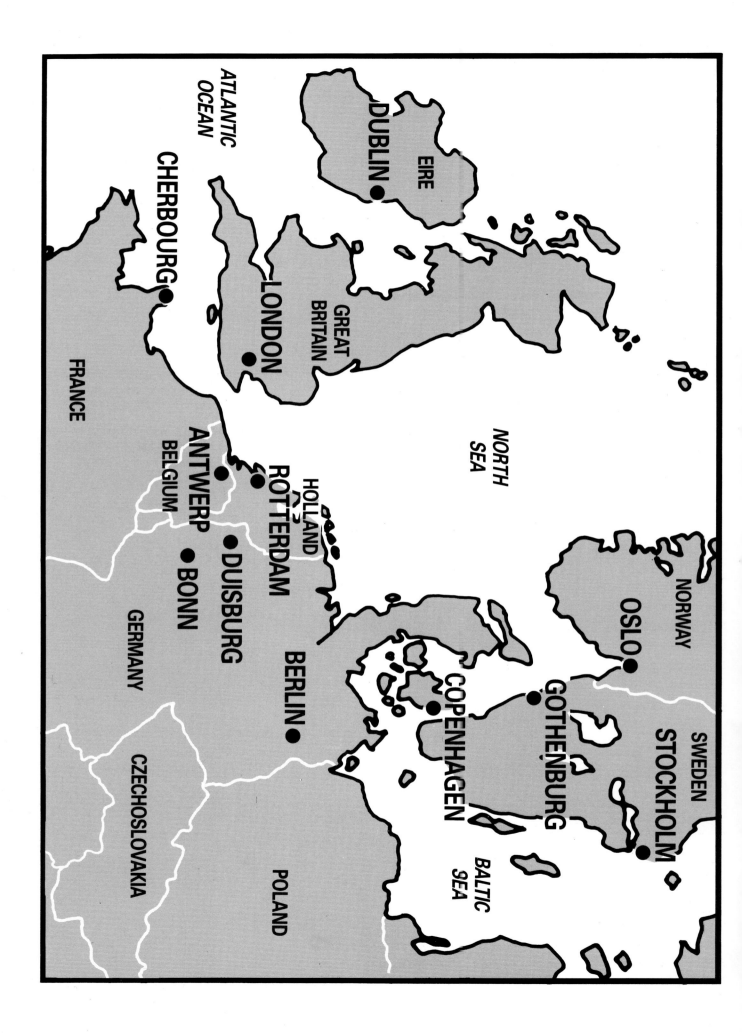

Contents

Preface
Belgium
Port of Antwerp

Denmark
Port of Copenhagen

France
Port of Cherbourg

Holland
Port of Rotterdam

Sweden
Port of Gothenburg

West Germany
Port of Duisburg-Ruhrort

Index

Acknowledgements

The Port of Antwerp early 20th Century, based on a painting by Eugeen Van Mieghem.

Belgium

Port of Antwerp

Mr A Himler and
Mr G Thues

Origins:
13th Century Creeks

The history of the port consists of a series of expansion programmes each on a larger scale than the previous one, for the construction of berths and installations.

The constant increase in the dimensions of ships has required an infrastructure on an increasingly larger scale: larger and deeper locks and docks, more spacious quay sites and sheds, higher and stronger cranes, as well as better facilities for road and rail traffic. The result of all this has been that the port's thriving attraction for more modern vessels and as a place to site every kind of industry.

The origins of port activity in Antwerp are based on the long concave bend in the right bank of the river Scheldt. At this location there was a small spit of land, which disappeared in 1885, about 0.2 hectares in size. It was called An 't werp (At the wharf), later Werf, from which the city probably took its name. Another name for this spit was 'the Crane Head' because there, from about 1263 until 1811, stood the sole and impressive treadmill crane which the city possessed, renewed in 1369 and 1546. The one from 1546 was a remarkable piece of engineering and its silhouette was known as far as Spain as a synonym for Antwerp. It was used for loading and unloading cargo, principally wine barrels.

In the period 1200-1250 three primitive inner harbours came into being which were little more than broad ditches. By 1568 their number had doubled as the result of the northwards expansion of the city in 1543-1545, during the 'Golden' 16th century. These six creeks opened directly into the river and were accessible only to small vessels which were left high and dry on the ebbtide.

First Docks (1811-1812)

When Napoleon paid his first visit to Antwerp on 18-21 July 1803 the situation in the port had remained unchanged for over two centuries. Because of its strategic location ('a pistol aimed at England's heart') the French Emperor issued two decrees which formed the basis of the new shape which the port was to take : a dock complex with a constant water level thanks to a lock. The decree of 21 July 1803 provided for the construction of an arsenal and of a military shipyard. The decree of 26 July 1803 ordered the construction of a dock with a lock and a 1,500m quay along the bank of the Scheldt. The spit of land at the river bend was retained so that an artillery battery could be positioned there if required. Two docks were excavated down river within the city walls. In view of their required surface areas of 2.59 and 7.13 hectares respectively, in the years 1804 to 1806 some 1,300 houses had to be demolished. The first dock with lock gates was officially inaugurated on 1 January 1811. This was done by the 'Friedland', an 80 gun warship which had been launched at Antwerp on 2 May 1810.

The access channel of this lock (filled in 1974) was 17.4m wide and the sill was a tidal lock with twin hinged gates at each end. These were opened at about two hours before highwater and closed about the same time after. The dock was 148x175m and ships were allowed with a draught of up 6.18m.

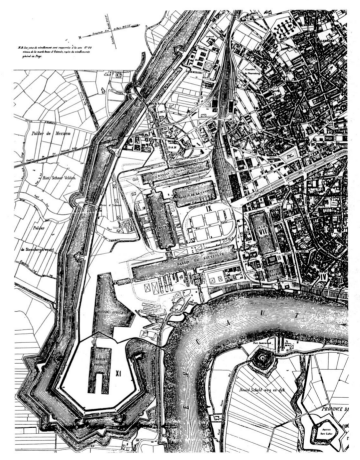

1 *The Port of Antwerp 1886. Map of the old docks north of the city and to the east of the river Scheldt. (Stad Antwerpen).*

There were only two quay walls and the other sides were slopes (scree banks). The work on the second dock began in 1808. When this 1.74x402m dock was opened to shipping on 29 November 1812 it only had scree banks.

These docks were called the 'Small Dock' and the 'Big Dock' respectively. They were designed and constructed by the French Engineer J N Mengin.

In 1903, the hundredth anniversary of Napoleon's decree, these names were altered to the 'Bonaparte Dock' and the 'Willem Dock'. The latter because after the fall of Napoleon in 1815 Willem I of the United Kingdom of Belgium and the Netherlands had presented both docks to the City which had to complete and maintain them at its own expense. More quay walls were built as shipping increased in volume. By 1837 the work had been completed. Typical of the trend which is still making itself felt is the fact that the surface area of the quays soon turned out to be too small. For this reason the Willem Dock was decreased in size to its present 5.73 hectares by widening the quays during 1875-1878 the Godefrid Quay was increased to 18m and the Entrepot Quay to 24m and in 1885 the Napoleon Quay to 10m. The channel between both docks is spanned by the Nassau Bridge dating from 1912. It is the last of 15 such hydraulic swivel-bridges built between 1887 and 1929 which still has its original mechanism operated by hydraulic power. It is now preserved as an example of industrial archaeology. The previous bridge at that spot, built in 1864, had been manually operated until 1878. It had in turn replaced the 'Iron Bridge' (so-called as opposed to the wooden bridges over the mouths of the creeks) paid for by the Chamber of Commerce and inaugurated by Willem I on 17 October 1822.

2 *An aerial photograph looking east of the beautiful old docks, Bonaparte Dock (1811) and Willem Dock (1812) showing the entrance lock from the river Scheldt, 1974. (Stad Antwerpen).*

Second Docks (1860-1869)

After the toll which the Netherlands had levied on cargo traffic using the river Scheldt had in 1863 been bought off by Belgium, which had in the meantime become independent, shipping increased rapidly. In 1864 2,753 seagoing vessels called at the port and the cargo traffic amounted to 895,085 tonnes. By 1900 these figures had increased to 5,244 vessels and 8,142,468 tonnes of cargo. This was also the period of the rapid progress of steam-powered and often iron-built ships, which in 1872 exceeded the number of sailing ships calling at Antwerp.

Already between 1856 and 1860 a larger dock 140x500m had been built further down river outside the city walls. This was the middle section of the present Kattendijk Dock with access to the river via a 25m wide tidal lock. In 1863 a drydock was built 160m long and 24m wide, which was joined two years later by two smaller ones 12x72m and 10x50m. To these was added in 1864 the Mexico Dock (150x380m) which soon after changed its name to Timber Dock, because of the original purpose for which it was used, namely the timber traffic.

For almost 10 years these two dock complexes, each with its own lock, remained separate until permission was granted to demolish the city wall which lay between them. The Kattendijk Dock was then linked to the two Old Docks by being extended 220m southwards and by the construction of the Connecting Dock. A weather-beaten large tablet fixed to the Willem Bridge recalls the solemn inauguration on 10 October 1869.

A goods station had been opened nearby in 1843. It served the 'Iron Rhine', the railway line linking Antwerp and Cologne. In 1847 a shed was built for cargo arriving and departing by rail. At that time old buildings in the immediate vicinity were being used as warehouses, but many cargo-handling firms were also erecting new warehouses, most of them equipped with stables, smithy, waggon and sailmaking shops and offices. By 1872 the 'Association of Wharfingers' had twenty member firms with 545 working partners or 'wharfingers bosses'. In 1900 their numbers rose to 31. The goods station disappeared in 1876 and the large shed in 1878. But this area has still the biggest concentration of 19th century warehouses.

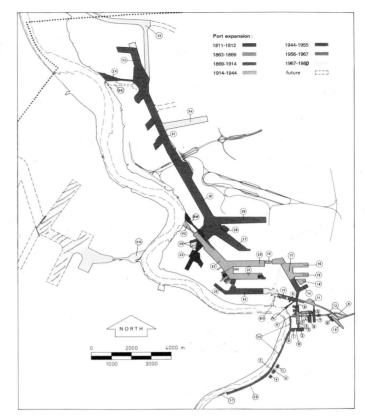

Port Expansion (1870-1914)

During the Franco-Prussian War of 1870-1871 Antwerp profited by Belgian neutrality, as a result of which 5,164 vessels carrying 2,449,233 tonnes of cargo arrived in the port in 1871. This trend continued to make itself felt later and required a series of new port expansion and improvement schemes.

In 1873 the Timber Dock was increased to an overall length of 500m and linked to two new docks, the Kempisch Dock (130x350m) and the Asia Dock (75x610m), which were excavated at the same time. The later dock was extended another 85m in 1888. The Kattendijk Dock had already been extended 320m northwards in 1881 and three dry docks 15m wide and 131m long were constructed. A special requirement was that up to 1873 all docks for seagoing vessels were rectangular and built in exactly N-S or E-W direction. For barge traffic the Old-Lobroek Dock (filled in 1957) was built during 1877-1879 (450x38m with a depth of 2.65m).

Over the period 1877-1888 the city river bank was straightened and a 3.5 km long quay wall built, as a result of which the spit of land (the Werf) and some 600 buildings disappeared. Cranes, iron sheds, numerous railway lines and a road were sited along the river in a 100m wide strip. Some of the iron sheds and cranes have recently been preserved as part of the port heritage.

About this time, the inner port had also been extended. In 1887 the America Dock (6.9 ha), the Lefebvre Dock (at first called the Africa Dock and since 1981 added to the America Dock, 10.7 ha) and their connection with the Kattendijk Dock was opened. The first of these together with the surrounding storage areas were reserved for oil, petroleum etc. In fact petroleum had been arriving in Antwerp since 1861, the first European port for this traffic. The oil arrived in timber barrels handled at a wooden jetty specially built for this purpose. From late 1863 onwards Antwerp was Europe's largest petroleum storage area.

3 *The sequence of dock development in the Port of Antwerp for the period 1811 to 1985. 1. Bonaparte Dock, 1811. 2. Willem Dock. 1812. 3. Kattendijk Dock 1863 (enlarged in 1869 and 1881).* **4.** *Mexico, later Timber Dock, 1864. 5. Connecting Dock, 1869.n* **6.** *Campine Dock, 1873-7. 7. Asia Dock, 1873. 8. Suez-cut 1887. 9. Africa, later Lefebvre Dock, 1887. (since 1981 incorporated in the Amerika Dock). 10. Amerika Dock, 1887. 11. Straatsburg Dock, 1935. 12. Albert Canal, 1935. 13. Lobroek Dock, 1938. 14. First Harbour Dock, 1907. 15. Second Harbour Dock. 1914. 16. Third Harbour Dock, 1914. 17. Albert Dock 1907 (enlarged in 1914 and 1928). 18. Shelter Dock for barges, 1922. 19. Connecting Gully, 1928. 20. Leopold Dock, 1928. 21. Fourth Harbour Dock, 1932. 22. Hansa Dock, 1928. 23. Marshall Dock, 1950. 24. Fifth Harbour Dock, 1960. 25. Industrial Dock, 1960. 26. Marshall Inset Dock, 1962. (enlarged in 1968). 27. Sixth Harbour Dock, 1964. 28. Grain Dock, 1964. 29. Churchill Dock, 1967. 30. Canal Dock B 1, 1967. 31. Canal Dock B 2, 1967. 32. Canal Dock B 3, 1967. 33. Scheldt-Rhine Canal, 1975. 34. Delwaide Dock, 1979. 35. Old Scheldt quays, 1882-1888 (quays 12-29). 36. New Scheldt quays, 1900-1903 (quays 1-11). 37. Oil jetty, 1905. (Stad Antwerpen).*

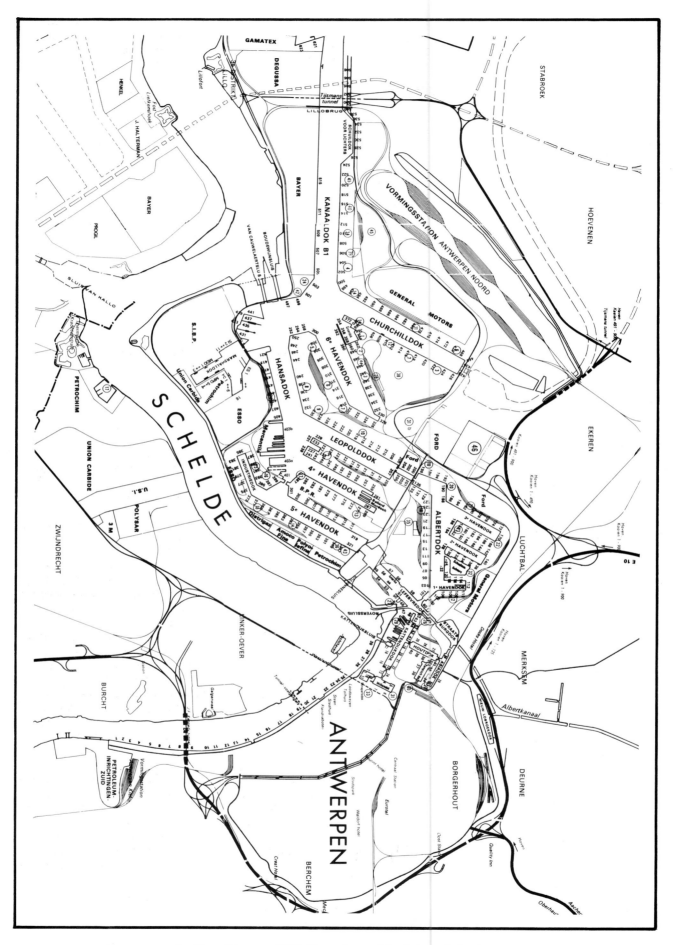

4   *Port of Antwerp 1987. General map of the thriving docks along the river Scheldt with numerous industries including Ford, Esso, General Motors, Union Carbide, etc. (Stad Antwerpen).*

It soon turned out to be a dangerous type of traffic requiring special safety precautions and isolation from other types of traffic. These precautions were further increased after the catastrophic explosion on 6 September 1889 in L. Corvilain's munition works, located more or less at the site of the present Royers Lock. This explosion took 84 lives while some 58,000 barrels of petroleum went up in the ensuing conflagration. For that reason in 1903-1904 the petroleum installations were transferred to a new 54 ha site with a 300m long jetty at the southern end of the Scheldt quays. The America Dock was then incorporated into the other docks.

4ᵐᵉ Section. — CANAL DES BRASSEURS.

5 *Old scenes at the Port of Antwerp.* **(a)** *A picturesque view of the river, warehouses and grain elevators in the 18th century.* **(b)** *Activities along the beautiful Brewery Canal with its fine warehouses during the second half of the 19th century. (Stad Antwerpen).*

Mechanisation and Hydraulic Power

Over the period 1870-1914 there was a gradual change-over from man and horse power to mechanisation. The four 10 to 40 tonne manual cranes of 1837-1867 were joined by more modern installations powered by water or electricity. Six 1.5 tonnes steampowered cranes were used from 1877 until 1885 and then sold as they suffered too much from sabotage.

The hydraulic installations were inaugurated on 24 November 1878. This was an English technological innovation from ca.1850, by Sir W G Armstrong (Newcastle-upon-Tyne, 1810-1900) which had proved its value in a number of large ports: London, Liverpool, Marseiles and Buenos Aires. Water was brought to a pressure of some 50 bars in a steam-powered station and carried via underground pipelines with a diameter of 150 mm to the hydraulic machinery of the many port installations linked to the network. At first these included the seven iron swivel bridges dating from c. 1867, ten twin lock gates, forty-three 1.5 tonne capstans (used mainly for hauling railway wagons) and eight 2.5 to 5 tonne capstans (at the Kattendijk Lock, used for moving the lock doors and hauling vessels rapidly through the lock) and the fixed hoist for loads of up to 120 tonnes. This latter, an enormous tripod, electrified to 550V in 1907, dominated the area of the Kattendijk Dock and the Eastern Quay until its demolition in 1935.

Between 1879 and 1912 these installations were joined by 335 1.5 or 2 tonnes hydraulic cranes mounted on rails. There were five kinds. When used at half of their reach they could serve as three tonne or four tonne cranes. The first ones were made in England and later in France. But from 1888 Belgian firms also supplied cranes and from 1901 they enjoyed a monopoly. In the St. Felix Warehouse (1863, industrial monument since 1976) simpler hydraulic lifting equipment was employed: two special English winches (0.4 and 0.6 tonne jiggers), three 1.2 tonne lifts and five 0.7 tonne bracket cranes. The latter were also used in a few other warehouses. Since 1975 the jiggers have been on show in the Iron shed near Steen Castle. The underground pipeline network was extended until it reached an overall length of c. 42 km and the most northern part of the America Dock.

The last thirty-four hydraulic cranes were taken out of service in 1975. Two typical examples are preserved at Scheldt (near Steen Castle where the National Maritime Museum is sheltered) as remarkable examples of technological skill dating from 1907 and 1912. When the hydraulic machinery of the swivel bridges was replaced in 1972-1976 by electrically operated and oil pressure driven machinery the last two hydraulic stations with steam engines and accummulators were taken out of service in 1976 and 1977, and are currently under preservation order.

Since the beginning of the 20th century it was clear that electrical energy was the thing of the future. In 1899 trials were carried out on an electrical capstan and in 1904 one hydraulic crane was converted to electricity. In 1906 the first 50 tonne electrical crane made its appearance at Scheldt Quay no.10, where it can still be seen. In 1907 eighty 2 tonne electrical cranes were purchased, to be followed in 1913 by thirty 2.5 tonners. These were the first two phases in a long series. By 1928 there were already 249.

Within one decade the new technology had become completely accepted. The stiff resistance marked by rioting in 1883 sparked off by three thousand dockers who considered that their livelihood was being threatened by the introduction of floating grain elevators was a thing of the past. Innovations frequently met with a negative reaction but since Antwerp wanted to maintain its position as a world port, the evolution could not be halted. In 1910-1914 ten floating pneumatic grain elevators were purchased from the German firm of A G Luther. These were steam-powered installations which could unload 180-200 tonnes per hour. The series was further expanded to reach a maximum of 24 in 1931. Two of them are still in use (No.17 and 22) albeit after some transformation and modernization. No.19 (1926) was sold in 1985 as a museum piece for Rotterdam and No.21 (1928) is preserved in the National Maritime Museum. Antwerp was also an important port for emigration: 1,086,153 people for 1843-1904 (last published statistics), most of them for New York: c.70%.

The period 1914-1944

Between the two World Wars port expansion continued with:
— the biggest municipal drydock no.7 (29.4m x 225m) built in 1919;
— the refuge dock for barges (9.6 ha) in 1922;
— the extension of the Albert Dock (now 70 ha), the construction of the Leopold Dock (53 ha) and Hansa Dock (247 ha) with the Kruisschans Lock (since 1962) called the Van Cauwelaert Lock, 35m x 270m, in 1928;
— the Straatsburg Dock (11 ha) in 1935;
— the new Lobroek Dock (10 ha) in 1939 to replace the dock of the same name excavated in 1879 and filled in 1957;
— Mercantile's two private dry docks (21.15 x 165m and 15 x 145m respectively) in 1930;
— the last municipal dry docks nos. 8 and 9 (both 20 x 151m) and 10 (15 x 100m) in 1931. They are no longer square to the quay wall but under 65 degrees, to facilitate the entrance.

The period 1944-1955

When Antwerp was liberated in September 1944 the port emerged virtually unscathed, which was an enormous advantage since after the Scheldt had been swept for mines the Allies were able to use the port to bring across large amounts of military supplies, which rightly earned the port the name of 'the port for the liberation of Europe'.

After the war traffic soon recovered and already in 1947 the pre-war record figures of 1937 had been exceeded. This naturally created problems. For example, ore carriers could not be unloaded within a reasonable period. Vessels wishing to enter or leave the port via the Kruisschans Lock, the only one large enough to handle them, 270m long, 35m wide and 14.01m deep, had to wait in a queue.

The City took note of the problems in due time and the necessary steps were taken to maintain Antwerp's international competitive position—before the war it had been the world's fourth port.
Thus the Marshall Dock (41 ha, at first called the Petroleum Dock) was opened in 1954 since already at this time the trek of industry from the interior towards waterways was making itself felt and two important refineries had expressed a desire to establish installations at Antwerp.

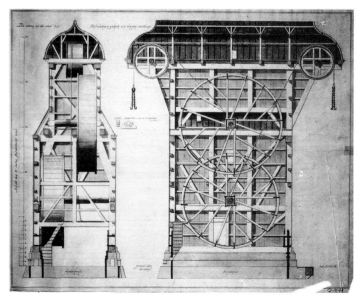

6 *The impressive double treadmill crane nearly 20m high, on the river Scheldt was constructed in 1546 and used until 1811 for loading and unloading cargo, principally wine barrels.*

Circa 1870

Circa 1914

7 *The period between 1811 to 1878 was marked by wooden sailing ships, some iron hand driven cranes, small quay walls and wooden sheds. From 1843 railtracks were constructed and rapidly extended. The photographs shows details of an iron hand driven crane. (Stad Antwerpen).*

8 *Between 1878 and 1914 the port view was dominated by iron vessels and iron sheds besides hydraulically driven cranes of 1.5 to 2 tonnes. **(a)** View of the Scheldt Afdah pitched roof sheds built around 1890 at the Houtdok (Timber Dock). Some of these sheds have now been restored to house a maritime museum. **(b)** Old port activities in 1925 outside the flat roof iron sheds built around 1903. (Stad Antwerpen).*

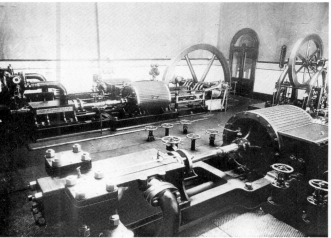

9 *The protected Gothic Style Hydraulic Power Station (Zuider pershuis) built 1882 with two accummulators in the towers.* **(a)** *View of the station 1986.* **(b)** *The steam engines replaced by electric pumps in 1959. (Stad Antwerpen).*

10 *The Port of Antwerp early 20th century.* **(a)** *An electric gantry crane of two tonnes capacity with a turning jib at work at Kattendijdok, 1924.* **(b)** *Busy shipping at the Albertdock with barges and lighters 1930. (Stad Antwerpen).*

11 *Surviving warehouses and wharves in the city of Antwerp.* **(a)** *Werf Uvlabnatie, currently used as workshops, 1977.* **(b)** *The Bell Warehouse has already been converted into luxury appartments and offices, 1982. (Stad Antwerpen).*

12 *St. Felix Warehouse near the centre of Antwerp was built in 1863 and was used until 1976 for the storage of tobacco, wool and wine. It is under a preservation order.* **(a)** *A view of the interior taken at the entrance, 1976.* **(b)** *A view of the first floor ceiling with timber beams supported on iron columns.* **(c)** *A magnificent view of the wine vaults, 1982. (Stad Antwerpen).*

For this purpose the Boudewijn Lock was built in 1951-1955 to relieve the Kruisschans Lock. Its locking length is 90m longer than that of its predecessor, it is also 10m wider and its water depth is greater by 50 cm. These new dimensions were the direct result of the ever increasing size of seagoing vessels, a trend which has continued since then to reach gigantic proportions.

The results made themselves felt immediately: long queues of vessels in the port and in the river disappeared. Moreover the second access to the port formed a safety valve in the sense that if one of the locks was out of order as the result of an accident or to carry out routine maintenance on the lock gates, from 1955 on a second lock was available.

For this reason the municipal authorities in conjunction with the government drew up a daring plan for expansion spread over a period of ten years. Because of the enormous scope of the work to be completed this has gone down in history as the 'Ten Year Plan'. Carried out over the period from 1956 to 1967 it doubled the facilities of the port in many important respects. The length of quay walls grew from 45km to over 90 km; the water surface area of the docks increased from 500 ha to 1,300 ha and the total surface area of the port zone (including land) increased to 10,000 ha.

Cargo-handling equipment too were adapted to meet the requirements of larger vessels: the reach, nominal load and speed of the quay crane cycle have increased. Whereas pre-war quay cranes had a nominal load capacity of 3 tonnes and a reach of at most 20 metres, since c. 1980 lifting capacities of 35 to 40 tons and reaches of 28 to 40 metres are by no means rare. In addition technical progress has made their working cycle shorter in time.

Ten Year Plan 1956-1967

The Marshall Dock and the Boudewijn Lock were two extremely useful additions to he wide range of facilities which Antwerp had to offer but there were not enough to ensure the future of the port. Its competitors were also making strenuous efforts and in addition the world's shipping traffic was evolving.

Situation in 1987

The large-scale Ten Year Plan did not put a stop to further developments to meet the requirements of today and of the future. Good port management requires a watchful eye to be kept on new trends and techniques in order to be able to take advantage of them in good time, bearing in mind that large constructions only come into service some four to five yars after the decision to go ahead has been taken.

Since the possibilities of expansion on the right bank of the river Scheldt are gradually becoming exhausted, everything necessary is being done to extend the port on the left bank of the river. The new complex will eventually be just as large in size and scope as the present port on the right bank. The initial port zone is around 6,000 ha.

The conclusion to be drawn from all of this is that as in the past the port is making every effort to remain competitive and modern, and provided that no obstacles are placed in its way from outside, Antwerp can look forward to the future of its port with confidence. The 80,000 or so people who earn their livelihood directly or indirectly from the activity in the port are one of the factors guaranteeing this.

The Port Today

At the end of 1989, the City of Antwerp Council launched an international competition to develop a plan for revitalising the former docks and redevelopment of the waterfront. The competition will be part of a master plan being developed for the whole city.

14   *Containerisation in the Port of Antwerp 1979. From the early 1960s extensive facilities were developed at Churchill Dock and Dock No. 6 for trade with USA and Europe. (Stad Antwerpen).*

13   *A view from the river of the preserved hydraulic and electric cranes and jibs as part of the National Maritime Museum established along the river Sheldt near Steen Castle. A collection of other items such as dockers' tools, ropes, anchors and engines are kept in the Iron Sheds at the rear of the cranes, 1982. (Stad Antwerpen).*

Acknowledgments

The sections of this article dealing with the period 1263 to 1944 were written by Mr Albert Himler and those for the period 1944 to 1987 by Mr George Thues.

We are very grateful to the City of Antwerp Council (Stad Antwerpen) for their kind permission to reproduce the photographs and illustrations.

15   *A panoramic view looking east from the modern docks of Insteek Dock west of the river Scheldt 1982. The aerial photograph shows the winding river with Marshall Dock, Churchill Dock, Haven Dock and Leopold Dock on the east bank. (Stad Antwerpen).*

Denmark

Port of Copenhagen

Dr Jorgen Sestoft

Historical Introduction

The port of Copenhagen was founded in the 12th century in the sheltered strait between Sjaelland and the small Strandholm, now Slotsholmen. The town emerged by the harbour, and the harbour gave its name to the town. Originally it was called Havn (Harbour), later it changed to Kobmannehavn (Merchant's Harbour). The geographical position made the town a competitor to the North German towns of the Hanseatic League in respect of the trade dominance in the Baltic.

The natural conditions for the development of the harbour were favourable. Neither water depths, bottom material nor sea currents caused severe technical hinderances. The harbour has normally only ice problems a couple of times a decade.

About 1500 AD the naval dockyard was founded at Bremerholm, north of Slotsholmen, where it remained until 1860. The first big harbour extension was made by the mercantilistic minded King Christian IV who moved the harbour to the stream between Sjaelland and the island of Amager. About 1600 he built a new naval harbour at Slotsholmen, and in 1618 a totally new fortified canal town, Christianshavn, was founded between Sjaelland and Amager. Christianshavn was built on dyked-in land based on Dutch design. In this connection the first bridge, Knippelsbro, was constructed between Copenhagen and Christianshavn, and King Christian IV had the Borsen (Stock Exchange) built adjacent to the bridge ramp.

At the end of the 17th century the fortresses north of Christianshavn were extended by artificial islands to form a new naval harbour, Holmen, which is still the naval headquarters. Copenhagen's fortification was simultaneously extended to the north to Kastellet (the Citadel), thus establishing a greatly increased and protected harbour area. The area, the limitations of which are still clearly seen on maps of Copenhagen, is today called the Old Harbour or the Inner Harbour.

In the 18th century the harbour was intensively used by the mercantilistic trading companies Asiatisk Kompagni, Vestindisk Kompagni, Gronlands Handelskompagni, and others engaged in the oversea trades. They were primarily sited on the Christianshavn side where there was also a good deal of industry and shipyards. The foundation in 1749 of the new quarter, Frederiksstaden, on the Copenhagen side (including Amalienborg Palace) gave a much needed extension of the harbour and quay areas. About 1780 a number of large company warehouses were built along this waterfront.

Copenhagen thrived as a trading metropolis until Denmark's unlucky participation in the Napoleonic wars. After 1807 business went into decline for the port of Copenhagen, which maintained its old size until the last decade of the 19th century. However, in the middle of the century an intensive domestic traffic with steamships brought new life to the harbour. A number of forts were built about 1860, and the company Burmeister & Wain moved its shipyard from Christianshavn to the new Refshaleo north of Holmen in 1872.

In 1891 a comprehensive extension of the harbour was initiated north of Kastellet. It was Copenhagen's Frihavn (Free Port of Copenhagen) that was built in order to enter into competition with Hamburg and the Kaiser Wilhelm Canal between the North Sea and the Baltic. This very ambitious project included docks with 9-10 metres depth of water, several warehouses and grain elevators. The most modern technology of that time was put to use, ferro-concrete as a building material and electricity for lighting and crane operation supplied by its own power station. In the Frihavn was also built a terminal for the train ferry service to Malmo, Sweden.

1 *Prospect of Copenhagen, 1611, seen from the east. Right of centre is the old naval yard existing until 1860. Left of centre is Slotsholmen with the old royal palace. In front of it is the rectangular naval harbour surrounded by the still existing arsenal buildings.*

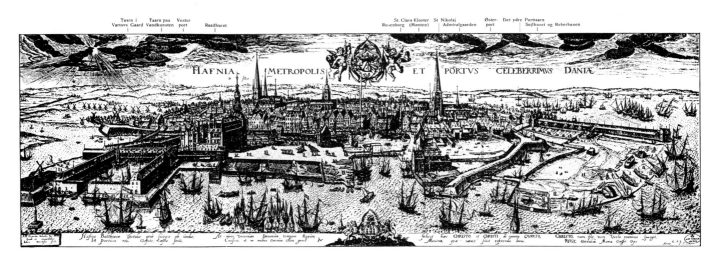

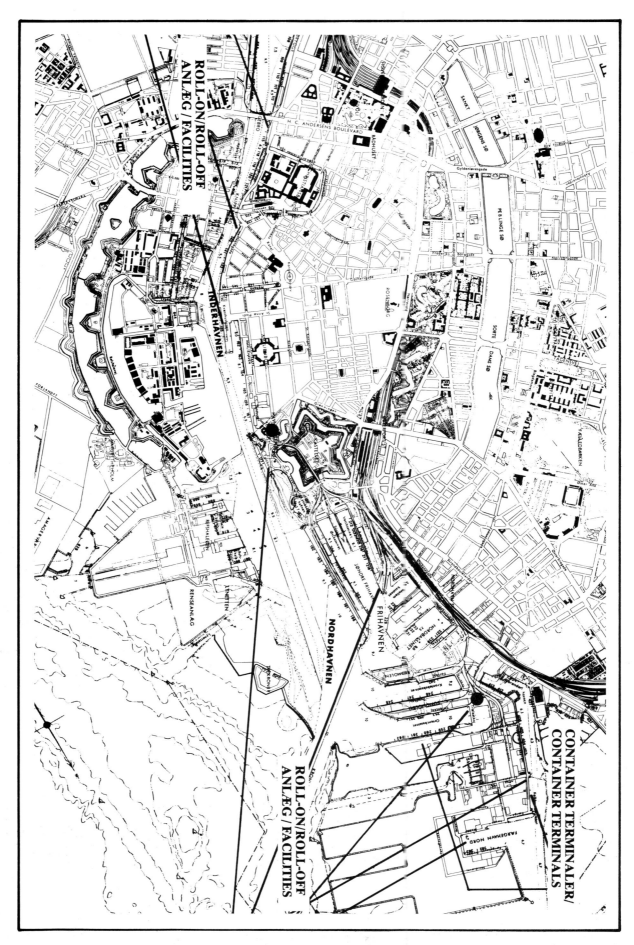

2 *Port of Copenhagen in 1988. General map of the docks and the Free Port in Denmark's beautiful capital. (Port of Copenhagen Authority)*

Through the coming decades the Frihavn became very important for the shipping lines between Copenhagen, America, and the Far East. As a transit harbour it became less important to the Baltic, because the Baltic countries' markets weakened after World War I

The Frihavn has by filling been extended several times to the north. The part of the construction that is best known internationally is the 900 m long Langelinie pier, Copenhagen's Quay d'Honneur.

From 1900 to 1925 the harbour south of the old harbour was extended. As it is only navigable from the north the access created traffic complications because of the two bridges crossing the fairway of the harbour. This part of the harbour was specifically used by harbour oriented industries. One of the old sea fortresses, Provestenen on the Oresund coast of Amager, was in 1934 turned into an oil harbour.

Since World War II the development has been concentrated to the Nordhavn where there are no problems with approach from the sea. However, the communications on land are not very good because the Nordhavn is only connected to the town by a small piece of land with traffic on both road and rail.

In the 1960s modern effective equipment for goods handling resulted in a decrease in harbour activities, even though the cargo turnover had increased. The container harbour in the north and the oil harbour in the east are out of reach of the general public. The traffic intensity has considerably decreased in the old harbour close to the town centre. The last 15 years have seen a further decline in the total goods trade, and many areas are now out of use or being used for other, non-harbour orientated purposes.

The Port Authority which owns and runs the harbour, is a private foundation with representatives from the private commercial sector and Copenhagen's council. The port of Copenhagen has today a total length of 42 km inclusive of canals.

Surviving Engineering and Architectural Features

Copenhagen's long waterfront is rich in historical and architectural landmarks. Buildings and dock equipment up to the 19th century are nearly all protected by law, whereas later sites are only protected in a few exceptional cases. In Denmark there is a public understanding that buildings and technical plant after 1850 may be worthy of protection and preservation is only in its first stage.

The old inner harbour within the rampart circle holds the largest number of protected buildings and plant, including the complete ramparts of Kastellet and Christianshavn, built in the 17th century by the military engineers Ruse and Hoffman respectively. Of the oldest naval yard on Bremerholm the magnificent gable of the anchor forge - now part of Holmen's Church - is still to be found, and so is a part of the long ropewalk, both dating from about 1560. The arsenal buildings at Christan IV's naval harbour on Slotsholmen are preserved, whereas the basin itself has been filled in. The Stock Exchange building dating from 1620 is likewise in existence, with its original use.

About 10 large company warehouses and several small, private warehouses dating from the 18th century have been preserved, while others have been demolished as a result of fire.

On the Christianshavn side the most impressive architectural feature is Asiatisk Kompagni's place with the main building built in 1738 by Philip de Lange and a warehouse from 1750 by Nicolai Eigtved. Since 1980 the building has become part of the Foreign Ministry's headquarters. On the quay side at Frederiksstad there are 5 warehouses from the 18th century, 4 of which are of the same design. The warehouses were built by C F Harsdorff. On the south side a pair of houses have been refurbished into a hotel, whilst the two adjacent buildings have been converted into flats. Another building on the north side houses casts of classical sculptures.

3 *View from Christianshavn over Slotsholmen, first part of 19th century. The spiral spire is the Stock Exchange dated 1620. The bridge to the right is the former Knippelsbro.*

4   Rigging shears and main guard on the naval yard Holmen,
1744-51, both by Philip de Lange. (Fotograf Jorgen Watz)

Immediately opposite the naval base at Holmen are several preserved monuments which include arsenal buildings, rigging shears, and a guardhouse by Philip de Lange, all dating back to the middle of the 18th century. From 1858 there remains a dry dock and a pump house by Henry Giles, a British engineer. Christianshavn's canal has a unique milieu with many fine private houses and warehouses. Between the canal and the harbour fairway is the Burmeister & Wain factory with buildings from the 1850s. Here is also the company's museum with models of ships and engines, emphasizing the develop-

5   Inner harbour looking north with Copenhagen to the left and Christianshavn to the right. On the Christianshavn side is the Foreign Ministry with new office blocks and adapted warehouses on the old site of the Asiatic Company. In the foreground the closed down engine works of Burmeister & Wain. The bridge is Knippelsbro dated 1937. (Fotograf Jorgen Watz)

ment of the diesel engine from the well known 'Selandia' in 1911. The assembly shop dating from 1923, by architect Niels Rosenkjaer is an imposing piece of industrial architecture. Close by the engine factory is the bridge built after 1618 bearing the name Knippelsbro. The delicately shaped 2-leaf basculating bridge with two copper-clad towers is from 1937 and designed by Kaj Gottlob.

In the Frihavn, the buildings in the most southerly and oldest part of the harbour, are worthy of preservation. The original project from 1890 was designed and built by H C V Moller, chief engineer of the port and the earliest buildings were designed by Vilhelm Dahlerup, a brilliant architect of his time. An outstanding grain elevator dating from 1894 was demolished after a fire, and the rest of the area shows traces of dilapidation. A grain elevator dated 1903 by Fr. Levy is the only protected building. The quay walls show pioneer work in using ferro-concrete in Denmark, and the four large warehouses at the Langelinie pier, 1893-1918, are also built with ferro-concrete structures. Some of these, especially Dahlerup's 4 storey warehouse of 1893, are of architectural and technical interest. They hold possibilities for re-use but are being threatened with demolition because of an ambitious hotel project.

At the port entrance the protected, but very dilapidated, fort Trekroner lies as a small independent island. Its present form dates back to the 1890s.

At the opposite end of the harbour - to the south - there is a lock and a construction for regulation of the water level, carried out during 1901-1903. The fine little lock keeper's house was designed by Dahlerup. In Sydhavn there were several interesting industrial buildings, but many of them have been demolished or modified when used for other purposes. Among the latter are shipbuilding halls dating from 1918 and Ford's assembly works from 1924. The power station H C Orstedsvaerket, which was constructed in 1916 and extended, has a historical technical monument, ie a Burmeister & Wain engine from 1932. With its 22500 horse power it was for decades the largest engine in the world.

6 *Warehouses on Christianshavn. To the right Asiatic Company's warehouse, 1750, now used for social purposes by the Foreign* *Ministry. The warehouse to the left, 1885, contains an Architectural Centre, exhibition halls and workshops for artists. (Jorgen Sestoft)*

Present Use

The Port of Copenhagen is today the local port of the town whereas earlier it functioned to a great extent as a transit port. The routing of the overseas traffic to the North Sea and Channel ports has meant a recession in business of the Port of Copenhagen. At the same time the domestic passenger and cargo traffic has changed into transport by road. There are no longer any harbour functions on most of the quays.

The Sydhavn has call for a few industries and silo plants, but usually it is used for laid up ships. From the inner harbour at Copenhagen City there are services to Malmo, Sweden, by hydrofoil vessels and regular service by large liners to Norway, Poland and the Danish island Bornholm. The canals are being used by yachts and veteran ships, and in Nyhavn the National Museum has two museum ships. The naval station Holmen is still in use and so is the naval dockyard. It is homeport too for naval vessels and various government and training ships. Burmeister & Wain's shipyard north of Holmen is still building bulk carriers and product tankers.

During the summer a large number of cruise liners come to the Langelinie pier where foreign naval visits also take place. Langelinie and the neighbouring park is a charming re-creational area attracting many visitors who especially come to see the small sculpture of the little mermaid.

The southern part of Frihavnen, west of Langelinie, is no longer in use. Some of the warehouses are hired out, some are empty. The modern container harbour at the Orient dock covers just under 10 hectares with considerable potential for extension and from where a feeder service to North European base harbours operates. In Nordhavn there is a deployment area for cars imported from Japan.

The latest innovation of November 1986 is a considerable increase in capacity of the goods traffic to Sweden, the so called Dan Link project with two four rail train ferries sailing between Copenhagen and Helsingborg. The turnover in goods is expected to increase by 2.5 million tonnes per year. The present turnover is about 7.5 million tonnes per year compared to a maximum of 13 million tonnes in 1970.

The Port Authorities have filled in many of the docks and leased them for commercial buildings with no connection to the harbour. They have sold some areas to private developers. This has compensated for losses in harbour business but has also caused criticism because the measures do not consider the public interest for a recreative use of the waterfront.

So far new housing has only been established on Christianshavn, whereas a number of big office blocks have been built near the harbour during the last couple of decades. Hardly any of these have aroused much enthusiasm, and the same is true with the official park just outside Amalieborg Palace paid for by a private donation.

Future Development

At present it is difficult to say anything for certain about the future of the harbour. There are many problems not yet resolved, as the planning for the future of Copenhagen is characterised by a mosaic of small local plans. The harbour function in itself has probably been stabilized after the period of decline and will remain to be concentrated in Nordhavn. This, however, calls for a long needed improvement of the approach facilities. There are also plans for moving the services to Norway, Poland, and Bornholm, from the inner harbour, to Sondre Frihavn (Southern Freeport) where it is possible to get better deployment area for cars by filling in the docks.

The international crisis for shipyards makes it uncertain as to how long Burmeister & Wain will exist, and it is also uncertain how long the navy will use Holmen. If this area becomes accessible to the public, Copenhagen may lose an attraction of great aesthetic, cultural and historical value. For reasons of protection and accessibility Holmen must be spared from intensive exploitation.

7 *Inner harbour seen from Northern Custom House. To the right one of the recent office blocks built for a shipping company, thereafter, 18th century warehouses. In the centre Port of Copenhagen's 150 tonne floating crane dating from 1962. In the background the two surviving passenger liners in the harbour, the Bornholm line and the Oslo line. (Fotograf Jorgen Watz)*

8 *Warehouse on Langelinie Quay, 1893, now out of use. The impressive building, by Dahlerup, has load bearing brickwalls and a Citadel from 1650. In the background the Nordhavn, where the container harbour is situated. (Fotograf Jorgen Watz)*

In 1985 there was an open competition for ideas for the Copenhagen waterfront, the result of which was published in a book (see bibliography). An obvious background for this was the 1984 competition for Oslo's waterfront. The competition for the Copenhagen waterfront gave a rich variety of projects more or less practicable, but their significance is uncertain as no authority is committed to use the projects. There is also a need for sponsors who are willing to invest in the many projects for concert halls and museums of modern art that appeared in the competition.

A positive benefit of the competition was the endeavour of the general public to have access to the water areas and recreational facilities such as parks, sea sport, bathing and other kinds of outdoor activities. However, a substantial barrier between the city and the water is not caused by the harbour itself but by the large railway areas, built at the turn of the century, north and south of the city. It is not possible to resolve this problem entirely as the improvement of the conditions will entail heavy expenditure.

Several competitors have suggested urban housing in the available dock areas. There is no doubt that there will be such a building development, but apart from actual plans for Christianshavn, it is hard to tell at the moment where else. The harbour administration is not averse to giving up areas of commercial interest even though they no longer serve the harbour.

It should be said that many of the proposed projects for the Copenhagen waterfront have borrowed ideas from the attractions of the existing historic areas instead of making the less pleasing places more attractive. This is the case for the prestigeous project of a congress hotel at Langelinie designed by the architect Jorn Utzon for a consortium of contractors. This project was carried out independently of the competition and was published later. It is supported by the Port Authority and the Lord Mayor but has many opponents. Its future is still uncertain. It is in competition with a similar hotel, project by the architects Henning Larsen and Bent Severin, located at Langebro at the northern end of the Sydhavn. This project can add to the quality of a rather desolate area and seems, all in all, to be more in accordance with the city plan than the Langelinie project.

Early in 1986 a competition was arranged for the re-use of the previously mentioned fort, Trekroner, which was an amusement resort but out of use since the second world war. Finally the state railways have plans for restoring a former Great Belt ferry, the 3000 tonnes 'Storebaelt' of 1951, and placing it in the harbour as a museum ship.

9 *Sondre Frihavn (Southern part of the Free Port). To the right Langelinie Quay. Bottom park and ramparts surrounding the* *inner structure with cast iron columns and reinforced concrete vaults.* *(Jorgen Sestoft)*

Acknowledgements

We are very grateful to the Port of Copenhagen Authority and the City Council for their kind supply of illustrations and photographs.

Bibliography

1. Akademiradet (ed) : Havnen og Byen, 1979.

2. K. Dirckinck-Holmfeld and C. Enevoldsen (ed): Ustopiens Havn, 1982.

3. C.Elling and V.Sten Miller: Holmens Bygningshistorie, 1932.

4. H. Fugl-Meyer: The Modern Port, 1957.

5. Arthur G.Hasso: Kobenhavns Frihavn 1894-1944, 1944.

6. Kobenhavrs Havnevaersen: Bidrag til Landshavneplanen, 1982.

7. Kobenhavrs Kommune: Kobenhavns Inderhavn, 1975.

8. G. Lorenz: Kobenhavns Havns Udvikling, 1934.

9. Nordisk Konkurrence om Kobenhavns Havn, Arkitekten, pp.317-347 1985.

10. Bent Rogind (ed):Den nordiske konkurrence om Kobenhavns Havn 1986.

11. Jorgen Sestoft: Arbejdets Bygninger, Danmarks Arkitektur, 1979.

12. Port of Copenhagen Authority, 7 Nordre Toldbod,DK 1259 Copenhagen K, has published various informative papers in English.

10   *Warehouse from l780s by C F Harsdorff before restoration and conversion to first class flats. (Jorgen Sestoft)*

France

Port of Cherbourg

Mr Alain Demangeon

Eighteenth Century Foundation

The idea of creating a naval shipyard-cum-large harbour on the French side of the Channel remained unchanged throughout the eighteenth century, and is attributed to one man, Vauban, who traced a harbour large enough to contain docks in front of the little seaport Cherbourg. Louis XV considered the project then dropped it when the English attacked the commercial port, but it reappeared once more in 1777, under Louis XVI. Cherbourg, like La Spezia in Italy and Carlscronn in Sweden, belongs to a very special family of harbours; this generation was born before steam and yet proved totally different from the huge 'cottage industry' type shipyards like Toulon, Rochefort and Brest, where most of the French warships were produced. Cherbourg could be said to be the first truly modern industrial port and certainly the most perfected of those constructed in France at the end of the Ancient Regime.

The reader may find an innovation like this surprising, when he considers how slowly industry got under way in France; it was years behind England, less concentrated and with a restraint that historians have always emphasized. There was slowness in making technical changes, absence of a specific architecture, and a late appearance of the forms of distribution characteristic of the industrial factory. The picture given does not compare well to that revealed by a study of the harbours at the end of the seventeenth century.

From a post-1773 position, this last generation of sailing shipyards obviously depended on events on the seas. The stability brought about after the Treaty of Paris, 1783, coupled with ambitions or trading imperatives, undoubtedly played an important part in the creation of the docks at Antwerp, Cherbourg or La Spezia which were to challenge England. Nevertheless, it must always be borne in mind that these three ports were not the only ones, but belonged to a series of projects for maritime sites which were drawn up between the the time of the French Revolution and Napoleon. They were not the characteristic sites of their time in the way La Rochelle was for example, but represented only part of a far vaster project to build a system of shipyards, docks, hospitals for sailors (lazarets) and defence harbours covering whole areas of the Continental coast.

And so it is that one cannot consider a port like Cherbourg as being a concept. Like all those dockyards, which never got beyond paper, and were thought of for sites like Calais, Saint-Malo or La Hougue, its history must be reconstructed within a movement which seems to have affected all the naval dockyards between the Revolution and the Empire. It can be said, in fact, that this system of dockyards and commercial ports fits into an approach to urbanism that is very characteristic of that of the late eighteenth century. It belongs to the same dream as occupying space in a rational manner, which prompted the modern State to conceive placing whole new towns wherever there were sufficient resources and prospects to justify them, or to consider the setting up of purely administrative towns as has happened in areas of France such as Vendee, Savoie and Brittany. This lead to the emergence of the idea of 'territory' as it was beginning to be considered at that time.

Construction of Dyke

No one had truly precise details concerning the respective qualities of the two harbours round which two geographically close rival projects, La Hougue and Cherbourg, had been situated in 1777. The Ministry of the Navy gave instructions that the first measurements be carried out.

Lacouldre de La Bretonniere (then Commander of the Vessels in the Channel) was commissioned to use sounding lines and to make a comparison between the two rival projects whose geographical situations were so close; Cherbourg won. He thought that the rather shoddy-looking harbour walls should be covered over by a 2000 tonne long pier. Closed off by this enormous sea wall, the harbour could have given protection to seventy big warships or, in other words, to the whole of the French Navy of the time.

The dyke or breakwater however, needed foundations twenty metres deep and several thousand metres long, and in Europe at that time nothing similar had been attempted. When La Bretonniere presented the findings of his mission, he suggested using 4000 ships, sunk offshore and filled with stones and pebbles, as the foundations for the dyke. The final decision was taken in favour of the Engineer from the Ponts et Chaussees school, Louis-Alexandre de Cessart, who had a very similar idea, consisting of closing off the harbour with 90 coniform oak hulls, 30 metres wide at the bottom and 20 at the top. They were to be linked together by steel chains and nets, and de Cessart claimed that they should 'decompose the effects of the sea and guard the entrance to the harbour'.

From 1783, only 18 of these remarkable structures were actually set up on the theoretical layout of the dyke, thus forming a series of anchorage spots that proved far weaker than expected. This solution was dropped after 1788, doubtless due to lack of funds and the accidental destruction of several cones. When the work on the infrastructure was finally completed in 1806, it had been done using loose rocks. Nearly 1000 men, mostly convicts or conscript peasants and then Spanish prisoners of war worked non-stop at this island of 3km circumference, which is still intact today off Cherbourg.

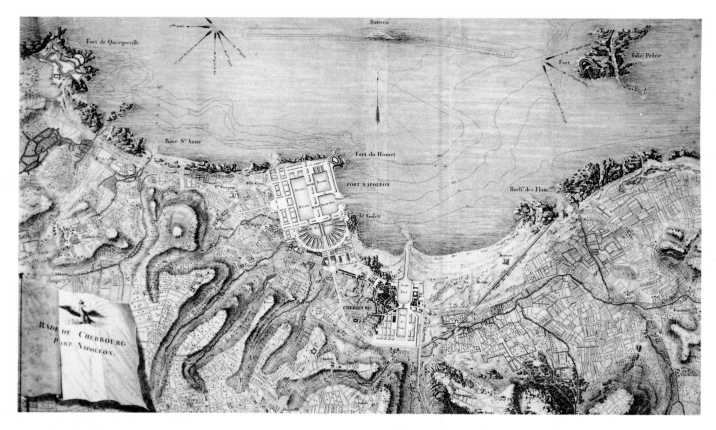

1 *Port of Cherbourg 1802. General map of the harbour with various forts and a battery on the centre of the dyke north of the port. (Musse De La Marine)*

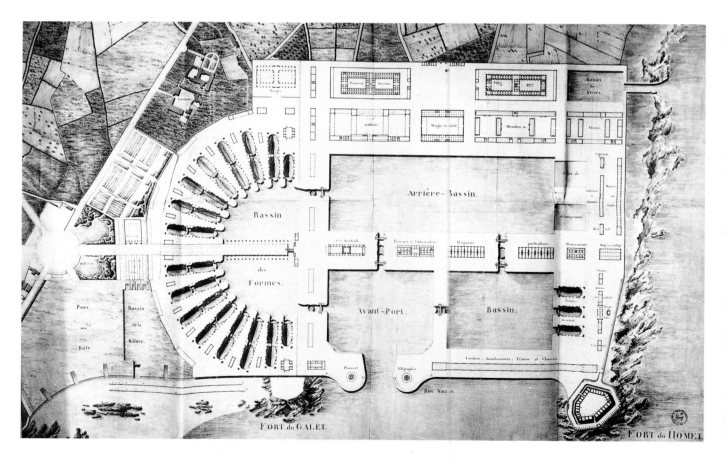

2 *Detailed plan of the docks designed by the Engineer Cachin 1802 and completed 1814, showing the outer dock, the inner basin and the graving dock for ship repairs. (Musse De La Marine)*

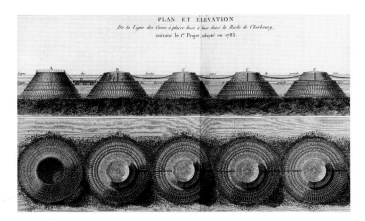

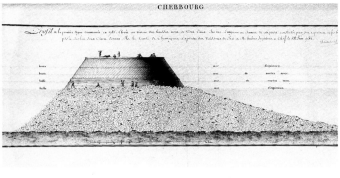

3 *Plan and elevation of the dyke or breakwater designed by the Engineer de Cessart in 1783. The cones were placed base to base in the tidalway of Cherbourg. Steel chains and nets were used for securing the cones together and forming a barrier to stop ships entering. The marvellous dyke is still intact today. (Bibl. Nat. Paris)*

4 *Cross section of Cherbourg Breakwater resting on the sea bed with stones piled around each cone. A battery of guns was mounted on the centre of the dyke, 1785. (Bibl. Nat. Paris).*

First the Harbour, then the Port

For twelve years, no real differentation was made between the docks and the harbour. However, after Thermidor, when plans were laid out to rebuild the French Navy thoroughly, Cherbourg was to be included in the huge network of military and industrial bases of which it was to remain, on paper at least, one of the most important.

On the 25 Germinal, Year XI, an official decree was issued, stating that Cherbourg should have a military harbour apart from that of the old town. 'I had decided to make Cherbourg like one of the wonders of Egypt' said Cachin in 1803.

The Engineer Cachin had a remarkable project to offer in answer to the demands of the Conseil des Travaux Maritimes;

1   The shipyards were no longer to remain behind fortified walls, but were instead to be like a town, open to the world where the restrictions of defence made way for a clearly productive logical composition.

2   The free distribution of the production units within the yards was coupled with a precise definition of what constituted the buildings and spaces; these were all different, but started from a set of rules in common.

3   Finally it can be seen that a central organisation of forms and slipways reappeared, inspired by the works of Thunberg for the Swedish port, Carlscronn, 1770. 'A semi-circular dock, 233 metres diameter, with open-access to the outer harbour. From the hub, its radii would terminate in fourteen double dry docks for inspection, rebuilding and refitting of warships.

Delorme and Forfait opposed the project, and so a tribunal was appointed by the minister to settle the dispute: Cachin was given support once more. Although the First Consul was for the widening of the programme ('a port able to shelter thirty big ships ready to go to sea at once'), he turned down the idea of the concentric shapes, preferring a rectangular inner dock. On 3 Vendemiaire XII, he gave his approval to 'the way the civil and military establishments at Port Bonapart planned to lay out the whole'.

On Napoleon's orders in 1804, 3000 Spanish prisoners and 2000 French soldiers started work on laying the docks, and this continued until 1813. Two steam machines were used non-stop to drain away the filtrage (only 20 or so existing in France then). Until 1814, Cachin pleaded tirelessly for the work on the infrastructures of the port (the fortifications, the jetties, the outer harbour and the western docks) to be pursued at the same speed as the construction of the civilian buildings in the docks.

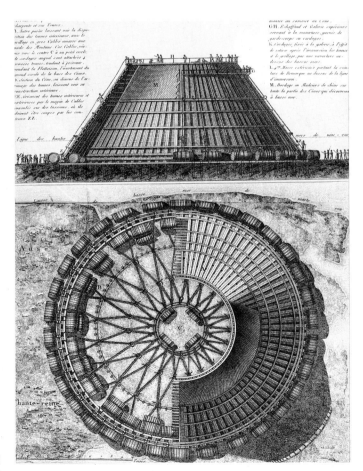

5 *Plan and elevation for the construction and flotation of the breakwater cone, showing the remarkable timber structure and the floatation barrels fixed to it by ropes. Rope ladders were fixed to climb to the summit of the cone. (Bibl. Nat. Paris)*

But since the Emperor's approval was necessary for every detail, the Maritime Works Council preferred to wait and gave their support only to the construction of the dry-docks and to those warehouses deemed indispensible for the building work. So it was that on 1st July 1813 the Empress opened the inner-harbour, a completely new space but one still without its industrial superstructures. On that day, the sea rushed into the 100,000 square metres of the docks into which all the Empire's funds had been sunk.

Cherbourg was very definitely there by 1814 and the inner-harbour had already been producing ships for two years. Yet this dockland would never be more than a huge workshop producing barely one tenth of the ships envisaged in the 1803 plan. Cachin would not relinquish the task however and work resumed after 1830. Although it was completed by the middle of the century, the only real function of the Cherbourg docks was producing steam warships.

Cherbourg therefore seems to have achieved what Rochefort, Brest and Toulon were unable to do; maintain its three docks, and the concentric line in its slipways and overall system, and be united by the buildings which constitute it. The composition of the docks can be said to well deserve our attention.

It was not that the Classical Age was unappreciative of the docks, ground and buildings as a whole but that a breakdown of the workspace with deliberate restructuring was required linked to the many work aspects anticipated. Organising the production of vessels in their entirety involved storage and maintenance of components, regular supervision and co-ordination of the different stages of assembly round the hulks and slipways and measures to combat fire and theft. Such organisation required precise definition of the use of ground space and a rational dispersion of workshops and assembly areas which do not have to be on the Lind docks.

Age of Ship Technology

Several explanations can be proffered; firstly that virgin sites had to be occupied very fast, that everything had to be thought of and put up in just a few years. This harbour urbanism was carried out after 1730 mainly by the engineers from Les Ponts. There would seem to be yet another reason for the shipyards, this being the rationalization of the ships themselves which played such a determining role in this transformation within the shipyards.

Jean Boudriot has studied the evolution which led the Navy to abandon the big sailing vessels of the classical period and to develop types of vessels which were faster, reliable and designed for 'mass' production. A whole century and two revolutions would go by (from Colbert's first instructions for the competition organised by Borda in 1783) before for ships were to be considered no longer as emblems or eternal prototypes of the French monarchy, but as products. This conception was transferred from the guilds to engineers and to the Marine Academy, which came into being at the same time as the first Ponts School, and those responsible were to experiment with the idea constantly. As a result of this change in outlook, ships were to be considered as war-machines and no longer as ancient monuments. The design was to be rationalized rather than individualised or embellished while the advent of production on a large scale was to raise the problem of the infrastructures and distribution of space necessary for making these ships.

Previously, the State had only questioned how the ships might react at sea; would they have sufficient speed and be easy to handle, could they accommodate up to 800 or 1000 men and fire the 74 or 118 cannons? Now, with most of the conception

6 *The towing out and putting in position of the breakwater cone by using boats pulled by oarsmen and possibly small steerage sail fixed to the top of the cone, l785. (Musse De La Marine)*

work being confined to one discipline, the 'models', and having at last achieved a 'type' from the prototype, the Navy allowed itself to be over optimistic. Repercussions from this unification of models followed immediately and it was presumed that since this worked with the models, those elements pertaining to the manufacture or defence of the vessels would be adhered to in one type. They were to be produced using uniform norms at the start and progressing to an ever-increasing size to the manufacture of prefabricated components which were put together round the construction hulls and in those sites which had become huge assembly yards. The standardization of the vessels brought about quite different results, however. Building costs decreased and by enabling the final product to be assembled in another place than where the components were produced, it gave rise to the establishment—later developed by the Council for Shipping during the Empire—of a network of shipyards and military harbours, maintained by the 'continental' factories. This hypothesis implied a relative relinquishing of the bit ports and a multiplication of bases in compliance with more restricted function (dry docks, assembly floors, defence, retreat etc.). This concept would later be applied to Belgium and to Holland, in part, with Antwerp, Terneuse, Cazdan, Breskens and Niev-Diep.

However, this dissociation of making and assembly of components brought up, in a new guise, the old problem of preserving ship maintenance. For many years this obligation to keep a sufficient number of vessels ready and waiting for battle without most of them ever going into battle, had been judged responsible for much of the 'shipyards' deficits. Too numerous and out of service, the ships became old, and required considerable maintenance, all of which doubled or tripled their price in just a few years.

The first idea put forward to overcome this problem was from Borda's circle; plan involving a certain number of 'flat' vessels for every port. These would not be in the water, they would not even be built, but would be kept quite simply in the warehouses, stacked in the correct order for assembly, and with all the sets of components available, which could be mounted on to the slipway 'at the first sign of war'. So it was that a vessel could hold good for ten years or more because its parts would have been made in advance. Thus instead of building ships which would deteriorate while not in use, we would merely store the component parts, and on the first command, a mere 3000 workers could construct six vessels in six weeks, which, had they been constructed gradually, would doubtless have entailed considerable maintenance costs'.

Whatever were the conclusions concerning these plans, it cannot be denied that Carlscronn in Sweden marked a turning point in modern shipyard technology. From then on emphasis was laid not only on the ships, the finished product, but also on how they were produced and how this was managed. It was with the ship as a starting-point that 'work' was analysed, and new demands were made: work procedure and stockage were rationalised, work phases on the site were better planned and there was more rigorous control of how it all advanced. All of which was to make itself felt quickly in that the projects at that time had to submit to far more stringent laws concerning use of space than previously. The evolution of the models took place, but the achievement seems rather to have been the conception of the efficient place of and for work. Cherbourg represents the acme of this change, for it was the only project of its time to have been carried out systematically. First in the creation of a totally man-made port protected by a 3000 metre dyke and then in laying the shipyards on a whole infrastructure of ramps, docks and slipways made out of solid granite. Such an abstract and visionary plan remains the only witness of this urbanism, which is surprising, where the spacing out and distribution of functions into autonomous sequences heralded another type of urbanity to come.

7 *Ship construction and technology was evolved considerably at Cherbourg with facilities and organisation for mass production. This* *illustration shows the launching of a new ship, late 18th century. (Musse De La Marine)*

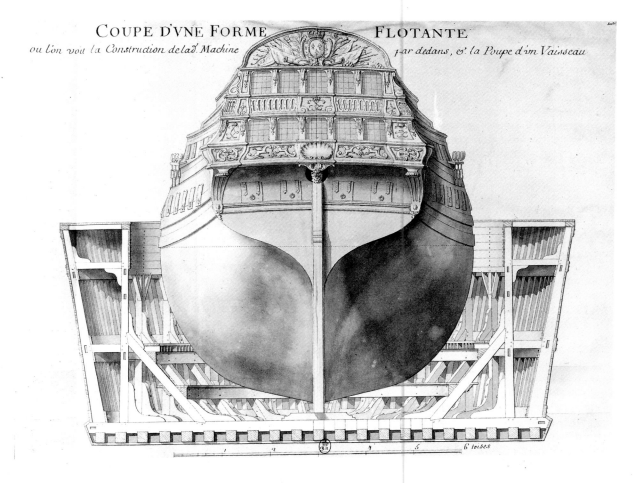

COUPE D'VNE FORME   FLOTANTE
ou l'on voit la Construction de la d. Machine   *par dedans, & la Poupe d'un Vaisseau*

The Port Today

After the Second World War, when it was completely destroyed, the port played its role in the setting up of the French strategic defence system. Although Cherbourg has separate modern commercial docks today, France's nuclear submarines are produced in the port and are launched from there.

Acknowledgment

Musse de la Marine and Bibliotheque Nationale Paris have kindly supplied the photographs and illustrations to whom we are deeply grateful for permission to reproduce.

Bibliography

1   In Le Vaisseau de 74 Canons, part Jean Boudriot, coll. Archeologie navale francaise, Paris-Genoble, 1974.

2   Scientifique, officier et ingenieur de la Marine, Borda fut l'un des fondateurs de l'Academie de Marine. Il fut charge, en 1775, de la reduction du nombrer type de vaisseaux de lignes, et de la coordination de leurs dessins entre les ports. Il est, avec Sane, l'organisateur des concours qui vont aboutir, apres 1780, au choix de trois types de vaisseaux de combat.

3   auteur avec bruno Fortier de l'ouvage 'Les Vaisseaux et Les Villes' Mardaga editeiu 1978.

8   *Cross section of the dry dock at Cherbourg with the stern of a ship under repair. The timber structure of the dock is clearly seen. (Bibl. Nat. Paris).*

9   *The Repualican Ship ran aground and broke up on the Rock of Mingant in Cherbourg 1794. (Musee de la Marine)*

Holland

Port of Rotterdam

Mr Han Meyer

Historical Introduction

Rotterdam belongs to the kind of cities which went through an explosion of growth at the end of the 19th century as a result of the industrial revolution. In the period 1880 to 1930 the population increased from 100,000 to 600,000 inhabitants. However, not only industry caused rapid growth, but also the strategic location of the city at the mouth of one of Europe's most important rivers, the Rhine, as a link between two nations with a fast developing industry: England and Germany. In later years it developed as a link between Europe and other continents; proudly Rotterdam gave itself the nickname Europort, the gateway to Europe. But now, at the end of the 20th century, the function of Rotterdam as a main port is not as self-evident as in earlier days. Rotterdam has to find new ways to survive as a city of international importance.

Until the middle of the 19th century Rotterdam was a city located on the right (north) bank of the river Nieuwe Maas. The city was situated on the edge of the impoldered peat moors of Holland. On the south side of the river was the river delta with its numerous islands and little polders. The riverbank of Rotterdam was the edge of the city and the face of the city to the sea and the world.

Originally the city subsisted on fishing, but after the Spanish conquest of Antwerp during the Dutch War of Independence (1564-1644), Amsterdam and Rotterdam took over the trade function of Antwerp. Amsterdam became the new main port of the Netherlands, but Rotterdam was able to develop a trade port with increasing interests of the West Indian and East Indian Trade Companies. Rotterdam had to construct new docks. At the end of the 17th century, next to the old 'landcity' behind the river dikes, Rotterdam had developed a 'watercity' by digging off and heightening the peat moors outside the dikes.

At the same time the city of Delft, with a strong manufacturing industry, tried to compete with Rotterdam by developing its own harbour: Delfshaven. This harbour soon fell into decay as a result of silting up problems and was annexed by Rotterdam during the 19th century. However, because of the complete destruction of the city centre of Rotterdam by German bombers during May 1940, Delfshaven is the only remains of early port development in Rotterdam.

1 *Map of the Port of Rotterdam in 1839 located on the north bank of the River Nienme Maas. (RPA)*

The extension of the trade function of Rotterdam made the city suited as a transit port between England and Germany. Already in the 18th century Rotterdam was the main port of Germany. Rotterdam owed this function especially to its strategic location at the mouth of the river Rhine.

At the same time Rotterdam was located at the north-south route of Amsterdam to Antwerp. The traffic on this route had to cross the river at Rotterdam by a ferry.

In the second half of the 19th century Rotterdam improved the infrastructure of east/west waterway and north/south overland routes. During the years 1866-1872 a new connection from Rotterdam to the sea was built: the New Waterway, designed by the Engineer Caland. This project made the port of Rotterdam accessible for large size ships.

2 *An aerial view of Rotterdam 1989. Right of the photograph is the south bank with the Kop van Zuid and left is the north bank with the city centre. (Aerocamera-Bart Hofeester)*

Five years later, in 1877, a new railroad connection was constructed between Rotterdam and Dordrecht-Antwerp. Next to the railroad bridge the Willemsbrug (Williams Bridge) was built for road traffic. This bridge was a result of an agreement between the municipallity of Rotterdam and the Rotterdamse HandelVereeniging (RHV, Rotterdam Trade Union), which was founded in 1872 by a number of influential entrepreneurs who wanted to develop new dock facilities on the left bank of the river. The RHV got the necessary lands on long lease, while the city authorities were bound to build the new bridge as necessary connections between Rotterdam and the new docklands. However, the ambitious RHV soon suffered financial problems and was forced to sell all their dock facilities to the city of Rotterdam in 1882. Since then, the development and exploitation of the docks has been the responsibility of the City of Rotterdam. In 1932 the City created a special Port Authority to fulfil this task at a professional level.

As a result of the new infrastructure of waterways and bridges, the port of Rotterdam developed rapidly. This growth took

place especially on the left bank of the river, adjacent to the crossroads of waterway and overland routes. The left bank of the river Nieuwe Maas was very suitable for the new extensions to the docks because of the special structure of the landscape. Like most of the cities of Holland, Rotterdam is dominated by the centuries old structures of polder dikes, ditches and water courses. In Rotterdam North the long polder dikes are dominant; they protect the peat moors with their narrow parcels of land from the river and the sea. South of Rotterdam is part of the conglomeration of ring polders and little islands, which dominate the whole delta area and have been connected together in the course of centuries.

The complete port area is situated outside of the main dikes and it is therefore separated from the systems of water regulation. Because of the fragmented structure of the left bank it is possible to change the course of the dikes in this area again and again, and to create new possibilities to the benefit of port development. On the north bank the extension of the docks is limited to the areas surrounding Delfshaven.

3 *Map of the Port of Rotterdam 1989. With its extensive international facilities, Rotterdam surpassed New York as world number one port in 1964 and has maintained this position ever since. (RPA)*

The Relation Between City and Docklands

In the old Watercity on the north bank, just as in Delfshaven, a strong mixture of shipping, harbour activities and city functions defined the relation between the city and the docks. The size of the ships and the character of the transhipment activities were such that until the end of the 19th century the street and quayside coincided. The increasing size of the ships during the 19th century required different kind of docks.

The first on the south bank shows greater independence from the city: larger docks, broader quaysides, big warehouses. Among these docks some vacant lands existed where it was possible to create small resident quarters.

In this part of the city the so called Kop van Zuid (head of the south), we find a fragmented complex of small and large docks, warehouses, islands, peninsulars, spits of land, mixed with small residential quarters, destined for dockworkers, seamen, immigrants, etc.

The Kop van Zuid developed in the period 1870-1910, is the last area with a strong mixture of city areas and docklands.

It is clear that the maintenance of this mixture is only possible because of their position outside of the dikes of the Kop van Zuid. The original 'watercity' on the north bank was completely destroyed by German bombers of 1940. During the reconstruction of the city in the 1950s the course of the main dike on the north bank was changed in such a manner that the watercity and Delfshaven changed to areas inside the dikes where dock activities became impossible.

The development of new docks after 1920 took place outside of the city to cater for the rapid increase in ship sizes and changes in cargo handling. During the 1930s the Waalhaven dock was built, being the largest basin in the world. The large scale development was caused especially by the growing importance of bulk cargo (oil, grain, ore, coals), which demanded large ships, broad quaysides and large storage sheds.

4 *An aerial photograph of the international Europort with the 'Maasulakte', an extension realized by impoldering part of the sea in the 1960s, is still lying fallow today. (RPA)*

From Transit to Distribution

Already during the 19th century the transhipment of bulk cargo dominated strongly the activities of the port of Rotterdam. However, the general cargo had regulated most of the employment, because of the high intensity of labour. So, the labour market of the port was defined strongly by the general cargo, but the ground use and layout for the most part by the bulk cargo.

Rijnhaven and Maashaven are docks especially designed to the benefit of bulk cargo handling, to make possible direct transhipment of the cargo from sea vessel into inland vessel, using special grain elevators.

After World War II the transhipment of bulk cargo was almost completely concentrated in the new dock extensions west of the village Pernis. With the tremendous extensions of the docks of Botlek and Europoort Rotteram was able to compete with Hamburg and to get into favour with the big oil companies which were undecided regarding their European locations between Hamburg and Rotterdam.

The construction of the large dock area Europoort during the 1950s turned the scale: a number of oil multinationals decided to settle in Rotterdam. Because of this development, Rotterdam surpassed New York as world number one port in 1964 and has maintained this position ever since.

A special aspect of the Europoort complex is the streaming of different kinds of traffic. The New Waterway is used by the passing through sea vessels; the Caland canal is used by the bulk cargo vessels with destination Europoort; and the low speed inland vessels normally use the Hartel canal.

Present Use and Future Development

The success of the Europoort brought in a new phase in the life of the port of Rotterdam. Because of the rapid growth of the oil industry, emphasis concerning bulk cargo has changed from transit to distribution. It is no longer just transhipment of oil, but also storage and production of different oil products.

At present a similar change is taking place concerning the general cargo and container handling. In 1967 the development of container transhipment was advanced by the establishment of the Europe Container Terminus (ECT): a joint venture of five important dock companies and the Dutch Railroad company. During the 1970s the ECT developed the biggest container transhipment company of the world, handling one million containers per year.

The change of use to a distribution port is connected with the changes in European economy. The significance of the Ruhr area as an industrial centre is declining, whereas the economy in the south of Germany and Switzerland is improving because of the growth of so called 'foot loose' industries. The consequence is that Rotterdam is losing the benefit of its geographical position. The river Rhine is no longer the natural connection with the hinterland. Nowadays Rotterdam tries to develop as a general distribution centre for the whole European continent and to improve the infrastructure of its railroad connections with the middle and southern parts of Germany and with France. The air traffic is improved by extending the small Rotterdam Airport.

5 *Rotterdam South Bank. A mixture of residential quarters and docklands on the Kop van Zuid 1986. (Aerocamera Bart Hofmeester)*

6 *The Free Entrepot. A large toll-free nineteenth century warehouse with 30,000 m2 area currently unused and awaiting redevelopment 1988. (RPA)*

7 *Declining activities in the docklands of the Kop van Zuid. (RPA)*

Surviving Engineering and Architectural Features

Another important aspect of the new economic policy of the city of Rotterdam is to extend the separate economic development of the city. In the future the port will be no longer the only economic strength of Rotterdam. The city is trying to attract new industries to create a new kind of economic life. In this respect tourism and recreation are given top priority in developing an attractive city.

Naturally, the old docklands will play an important role, because of their magnificent position on the banks of the river. They are considered to be features of the new attractive city.

During the second half of the 1960s some politicians became too confident of the success of Europoort and wanted to extend the port to gigantic proportions. A new extension was realized by impoldering a part of the sea: the Maasvlakte especially destined for heavy industries like blast furnaces, nuclear reactors, etc. Because of protest movements, these kinds of industries never materialized. Not only because of protest movements but also because of economic changes, large plots of the Maasvlakte are still lying fallow today. Nevertheless the city administration wanted to go ahead. In 1968 the proposal 'Rotterdam 2000' was presented. According to this plan the islands Voorne Putten and Goeree Overflakkee at the south side of Rotterdam, would have been changed into an industrial city of docklands. But the port of Rotterdam had already reached the limits of its growth. 'Rotterdam 2000' was not accepted by the City Council and the plans were scrapped.

Today, Rotterdam has a total length of quays of 40 km with a water area of 2148 ha. It handles 30,000 ocean going ships and 170,000 river barges every year with a total sea borne goods traffic of 270 million tonnes.

Quayside Construction—The soft and weak subsoil in Rotterdam demands special constructions. Until 1915 the quayside foundations were composed of wooden piles; from 1915 until 1955 predominantly concrete caisson quaysides were used. After 1955 the practice changed to steel damwalls, filled up with ferro-concrete.

Warehouses and Other Dock Buildings—The different kinds of dock activities have left behind different kinds of architecture. Early port developments are to be seen in Delfshaven; this little 17th century harbour has remained as one of the most picturesque parts of Rotterdam.

There are large warehouses for general cargo from the 19th century. A beautiful example is the Free Entrepot building constructed in 1879 by the RHV (architect unkown). This warehouse has a length of 200 meters by a breadth of 37 meters and has five floors with an area totalling 30,000 square meters.

The passenger shipping has flourished in the port especially on the Wilhelminaquay and has left behind some remarkable buildings. The office building of the Holland America Line (HAL) was built in 1901 to the design of L V D Tak. The arrival hall of HAL was added in 1949, (Architect Brinkman, van de Broek & Bakema).

8 *The Peperklip is a residential super block on the former quayside. The Architect was Carel Weeber, 1980. (RPA)*

9 *Maas Tunnel, built 1938-41, was the only river crossing tunnel in Netherlands for a long period of time. The photograph shows the ventilation building and entrance for cyclists and pedestrians 1987. (RPA)*

10 *Rotterdam Europort. Concerte windscreens for protecting low speed sea vessels against strong south westerly winds 1982. (RPA)*

A special product of the thirties is the HAKA Building (1932, architect Mertens & Koeman), which is used for several functions: storage, offices and workshops.

The increasing importance of bulk cargo has left a number of impressive silo buildings; for example the grain silo Brielselaan (1901, Architect Stok, extension in 1931 by Architects Brinkman & van der Vlugt), and the silo complex of the GEM (Grain Elevator Company) in the Botlek (1965 Architects Croese, Postma & Postma)

Dry Docks—The complex of the RDM. (Rotterdam Drydock Company) dates from the beginning of the 20th century (1903-1935, several architects). Adjacent to this site is the garden city Heyplaat which was founded in 1917, especially designed for the RDM workers. Nowadays because of the decrease of the ship building industry, Heyplaat is a quarter with the highest rate of unemployment in Rotterdam.

A more recent example of dry dock architecture is Van der Giessen-de Noord with the largest covered dry dock in the world (1982 Architect W Quist).

Bridges and Tunnels—Bridges and tunnels have always played an important role in Rotterdam. It has 440 bridges and three river crossing tunnels. The railroad lifting bridge, constructed in 1927 and designed by the Engineer Joosting, has been developed as a symbol of the connection between the north and the south bank. Unfortunately this bridge will be demolished because of the construction of a new railroad tunnel under the river. The new Willemsbrug (Williams bridge) is also a remarkable part of the skyline of Rotterdam.

For a long time the Maas tunnel (1938-1941, designed by the Department of Public Works) was the only river crossing tunnel in the Netherlands. For this reason the tunnel was and still is a feature of Rotterdam's show piece and indeed, its jewel.

Regeneration of Old Docklands

During the 1970s the economic changes were to be felt in the dockland activities. The former strong general cargo sector and the ship building industry were confronted with a shrinking market, reorganizations and closures. In 1976 the City Council presented the first proposals for reconstruction of the old docklands in the city. The proposals had two aims: first, the reorganization of dock companies, for more efficient use of available docks and second, the regeneration of the former

docklands into housing etc. During the decade 1976-1986 a large part of the old docklands were redeveloped. Small dock companies were amalgamated at suitable sites west of the city. The vacant areas of the old docklands were converted for new uses. In 1979 the city of Rotterdam concluded an agreement with the national government concerning the development of 8000 dwellings on former docklands, with a differentiation of 60% social housing and 40% free market housing. Today, the major part of this programme is completed or under construction.

In the old dockland areas near the city centre not only houses are being planned but also a mixture of business activities is generated. A noteworthy project is the Oude Haven (Old Harbour), designed by Architect Piet Blom and built in the early 1980s, radiating a mediterranean and holiday like atmosphere.

This project is part of the former Watercity bounding the central business district. The whole Watercity has got a destination as a new centre for culture, recreation and tourism. As part of this programme the new Naval Museum was built in 1986; (Architect W Quist), including an outdoor museum of old ships.

Elsewhere in the Watercity new office buildings, museums, appartment buildings, hotels, are under construction. Other old docklands are filled up predominantly with housing. Architects and urban planners have to deal with the large scale dimension of the docklands. The most remarkable product of this period is the housing project Peperklip of 1983. (Architect Carel Weeber), consisting of 549 social apartments in one super block, with the dimensions of a super tanker. This project has played a pioneering role in the reconstruction of the Kop van Zuid, which nowadays is one of the most interesting areas of urban regeneration.

During the 20th century, Rotterdam has been developed into a double city divided by a river with a port area. Rotterdam South has always been a subordinate part of Rotterdam: it was and still is the less priviliged part of the city with a lack of some services and inadequate connections with the city centre, etc. The new masterplan for the Kop van Zuid 1987, designed by T Koohaas, tries to remedy this position of the south bank. There is a plan for the Wilhelminaway as an extension of the city centre. On the left border will rise a Little Manhattan with a mixture of offices, residential functions and culture and recreation. This metropolitan function of the Wilhelminaway is the answer to the lack of services of Rotterdam South.

The Rijnhaven, south of the Wilhelminaway, will maintain a part of its port function. It will be developed as a floating city a service for the large number of inland vessels which have to wait for their cargo handling.

The most important element of the proposed infrastructure in the Kop van Zuid is a new bridge, which will connect the centre of Rotterdam with the Wilhelminaway. This bridge will become the key to develop a new main traffic axis in Rotterdam: Rotterdam Avenue connecting Rotterdam Airport in the north of the city with the Zuidplein (south square) in the south. The project is meant to inject new economic life in this part of Rotterdam, where the unemployment in some quarters has risen up to 40%.

When all plans are completed maybe Rotterdam will no longer be a divided city, but a united port with the river in the middle as an attractive urbanistic cultural and recreational focus.

11 *The 'Oude Haven' (Old Harbour) has been transformed into a beautiful residential and recreational area 1988. (RPA)*

12 *Rotterdam city centre. New building activities along the waterfront 1988. (RPA)*

Acknowledgment

We are deeply grateful to the City of Rotterdam Council and Rotterdam Port Authority (RPA) for their kind permission to reproduce photographs and illustrations. The aerial photograph of Port of Rotterdam was kindly supplied by Aerocamera Bart Hofmester.

Bibliography

1 "Rotherdam in de 19e en 20e eeuw", L J C van Ravesteijn, Leiden 1949.

2 "Rotterdam, stad in beweging". R Blijstra, Rotterdam 1965.

3 "Staf van Zaken, 50 jaar Scheepvaartvereniging Zuid", Rotterdam 1957.

4 "De haven van Rotterdam", J Ph Bax, Rotterdam 1929.

5 "Rotterdam verstedelijjkt landschap", Frits Palmboom, Rotterdam 1987.

6 "Mensen maken een stad", J Nieuwenhuis, Rotterdam 1955.

7 "Gemeentewerken Rotterdam 1955-1985" several writers, Rotterdam 1985.

8 "Yearbook Rotterdam Europoort", Rotterdam 1987.

9 "Havenarchitektuur", Rotterdamse Kunststichting, Rotterdam 1982.

10 "De Kop van Zuid", ontwerp en onderzoek", Rotterdam Kunststichting 1982.

11 "De Kop van Zuid, een stedebouwkundig ontwerp", DROS Rotterdam 1987.

12 "Rotterdam Waterstad", DROS Rotterdam 1987.

13 "Herstrukturering Oude Havens", Gemeente Rotterdam 1976.

13   *The Kop van Zuid is a mixture of residential quarters and docklands. The 'Peperklip' a superblock built on the former quayside, is seen in the centre of the picture 1988. (RPA)*

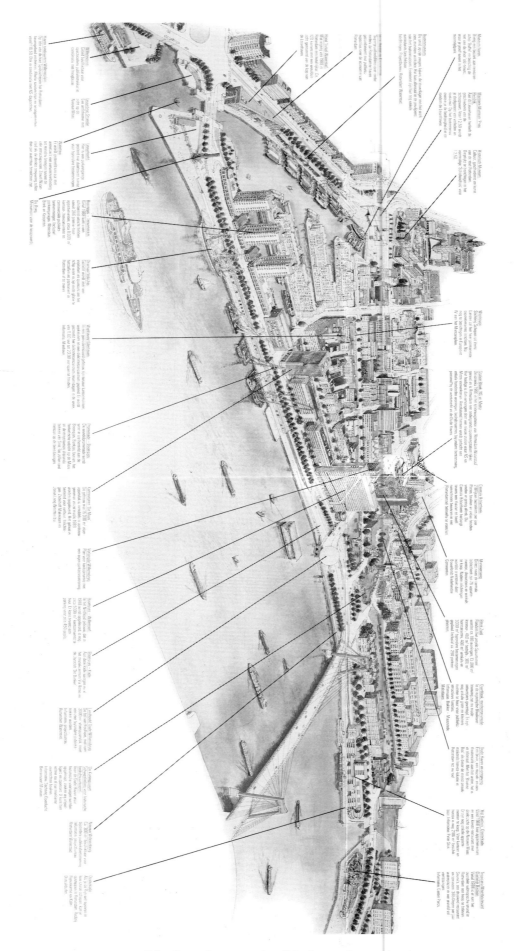

14 *Rotterdam watercity of the future. An artist's impression showing the proposed office and residential blocks, museums, marinas and other recreational facilities 1989. (RPA)*

Sweden

Port of Gothenburg

Mrs Gudrun Lonnroth

Early History

Gothenburg, with a population of approximately 430,000, is the second largest city in Sweden. The harbour, which has approximately 20 km of quays is the biggest in Scandinavia.

The Old Town which today is the city centre, is situated on the south bank of the River Gota. From the Old Town there is a relatively narrow area of docks, approximately 4 km long, extending westwards. The harbour and dock areas on the north bank of the river are broader and continue, with some breaks, as far as the coast.

Since the Middle Ages the River Gota has been of great significance to Swedish trade to the west. The west coast of Sweden at that time belonged to Denmark-Norway but in the 13th Century Sweden succeeded in forcing a passage along the river down to the coast. In 1658 the west coast became totally Swedish.

The oldest Swedish town along the River Gota was Old Lodose and was situated quite a long way inland. In 1473 New Lodose was founded at a location east of the present City of Gothenburg and it took another couple of attempts to expand further out towards the coast before the present City of Gothenburg was founded in 1619 on the south bank of the river, approximately 5 km from the estuary. During the 17th and 18th centuries Gothenburg developed into a flourishing commercial centre due mainly to the export of iron and timber from the upper reaches of the River Gota. The Swedish East India Company also played an important role in the development when it ran its operations from here from 1731 to 1813.

Gothenburg was built as a fortified trading town based on the Dutch pattern of canals, moats and walls. Because the inner docks, the Stora Hamnkanalen (the Large Dock Canal), were relatively shallow, trading ships anchored in the river outside the town or further westwards at Klippan, where the East India Company had a maintenance dock. Stora Hamnkanalen became the centre of trade. Here the East India Company had its warehouses and offices, the town customs house was located here as were all the most prominent trading houses.

Along the banks of the river, between the town and the Klippan Dock in the west, a suburb, Majorna, grew up. Here jetties and storage depots, shipyards and rope works were built alongside a dense residential development.

In 1807 Gothenburg ceased to be a fortified town. The demolition of the walls was carried out during the next few decades, but the town and the port still retained their original character right up to the 1850s.

1 *Map showing Gothenburg and its surroundings in 1844. On the south bank you can see the old town with the first quays along the river and in the west the suburb called Majorna. Through the marsh areas on the north bank there is a passage for a ferry boat. Here the first bridge was built in 1874 and the Gota Alv Bridge in 1939.*

2   *The Large Dock Canal with small boats carrying goods to the
East India Company's building. (Etching after a drawing by E
Martin, 1787).*

3   *The old town looking south east about 1870. From the left:
Packhus Quay with a ship for emigrants in front of the City Customs
House, Stora Bommen Dock connected with the Large Dock Canal,
Stone Pier and Seppsbro Quay. (Lithography after a drawing by O A
Mankell).*

Port Development
(1850–1950)

From the middle of the 19th century maritime traffic through
Gothenburg increased significantly. Among other things this
was the result of the establishment of a canal system (Gota
Canal and the Trollhatte Canal) together with the River Gota
which finally linked the east and west coasts of Sweden. What
was now required were port facilities to meet this increase in
traffic.

A plan for dredging and the building of quays was drawn up in
1843 by the 'Royal Committee for the Work on the Port of
Gothenburg and the River'. The plan covered the building of
docks adjacent to the Old Town, which were completed in
1859 together with the plan for the extension of the railway
network along the quays.

At the beginning of this century the shipyards, along with two
large timber factories and two mills, occupied a large area of the
river banks.

In 1874 a plan was drawn up for the continued expansion on
the south bank of the river. The plan was prepared by J G

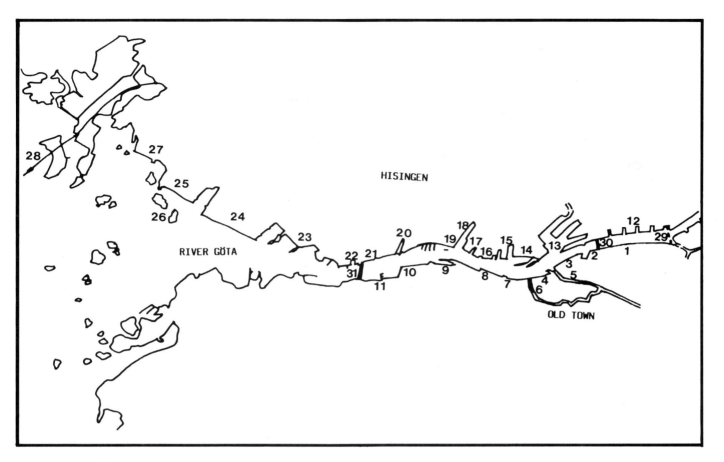

4 Map of Port of Gothenburg 1987
1. *Gullberg Quay.* 2. *Lilla Bommen.* 3. *Packhus Quay—Packhus Square.* 4. *Stora Bommen—Skeppsbro Quay.* 5. *The Large Dock Canal.* 6. *Rosenlund Quay, Fishall.* 7. *Masthugg Quay, Stigberg Quay (Stena Line).* 8. *'Ameica Shed'.* 9. *Fish Dock.* 10. *Majnabbe Dock (Stena Line).* 11. *Klippan, former East India Company's dock area.* 12. *Ringon (small docks, workshops etc.).* 13. *The Free Port and Lundby Dock.* 14. *Shipyard 'Cityvarvet', former Gotaverken.* 15. *Lindholmen Dock.* 16. *'Projekt Lindholmen', former Lindholmen shipyard.* 17. *Lindholkmen-Slottsberget residential area.* 18. *Sannegarden Dock.* 19. *Sorhallsberget.* 20. *Fomer Eriksberg Shipyard and the mill 'Tre Lejon'.* 21. *Farjenas, a ferryplace frolm the Middle Ages until 1967.* 22. *Rya Dock.* 23. *Skarvik Dock.* 24. *Scandia Dock.* 25. *Alvsborg Dock.* 26. *New Alvsborg fortress built in the 17th Century.* 27. *Arendal Shipyard.* 28. *Tor Dock.* 29. *The Tingstad Tunnel.* 30. *The Gota Alv Bridge.* 31. *The Alvsborg Bridge.*

Richert who was head of the City's bridges and dock buildings. According to the plan the expansion would have been both east and west of the Old Town. Meanwhile, in 1874, a bridge was built over the river which became an obstruction to development in the east. The most important extensions were therefore made westwards.

Around the turn of the present century there was a big debate as to how the Port of Gothenburg should develop in the future. One proposal, presented in 1892, was to build a free port and in 1904 an international dock plan competition was arranged. J G Richert won the first prize. His suggestion was based on the idea that a dock island with a central dock would be created out in the River facing the Old Town. The plan was strongly criticised and a compromise was reached which resulted in a decision in 1912 to construct docks adjacent to the marshy areas on the north bank.

Alongside this discussion, plans of lesser proportions were being drawn up for an expansion westwards along the river for the construction of Fish Dock and Stigberg quay.

In 1934 the City Council approved a general plan for the port. It proposed the continued expansion along the north bank and included a bridge that would link the city centre with the north bank. A proposal put forward in 1928 to build a tunnel under

the river had previously been rejected. The Gota Alv Bridge was finally completed in 1939.

Between 1850 and 1950 there was continuous expansion of the Port following the various plans put forward; first along the south bank next to the Old Town and also in the suburb of Majorna which was incorporated in 1868. After 1906 the north bank at Hisingen was annexed to Gothenburg and even here docks were built under the auspices of the City Council.

The conditions surrounding the expansion were very different for the two sides of the river. An important difference was that the docks on the south bank of the river grew adjacent to older residential areas and older industries whilst the north bank developed into an industrial and dock estate which was expanded in a more rural area and, for the most part, without any direct contact with residential areas.

The docks have mainly been created through dredging and infilling of, among other things, the large marsh areas which originally existed east of the City and along the north bank of the river.

From the 1890s most of the quays were equipped with electrically-driven cranes and dock sheds of timber construction.

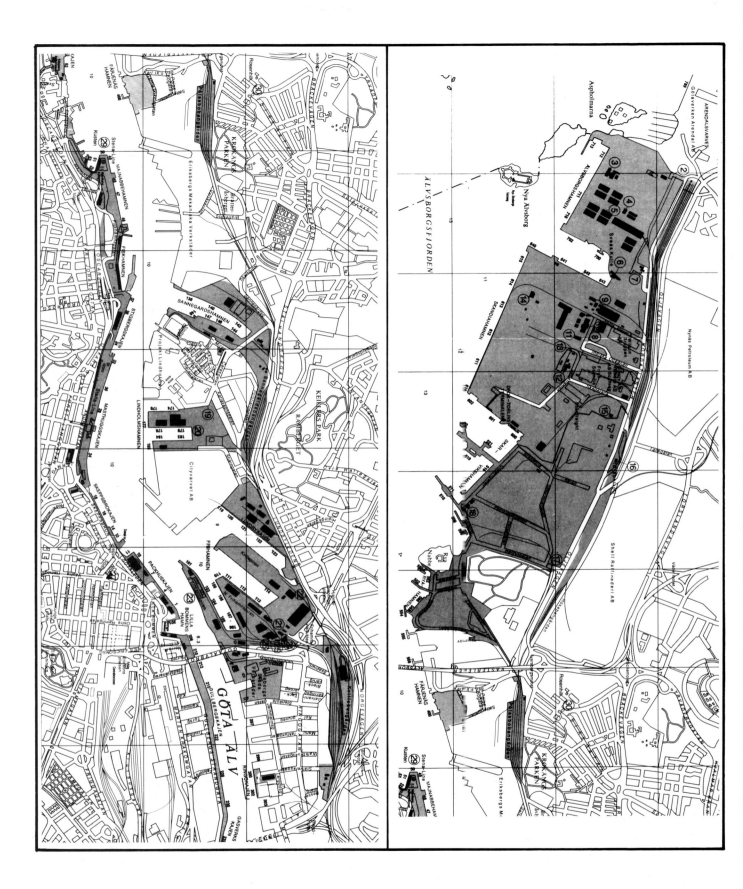

5   *The Port of Gothenburg today. Top: North Bank western part showing the Rya, Skarvik, Scandia and Alvsborg Docks. Bottom: The Old Town with the Free Port and the shipyard City Varvet on the North Bank and various quays and buildings on the South Bank. 1987. (Port of Gothenburg AB, B W bild).*

South Bank of River

The first stage in the expansion along the south bank of the river was completed around 1860. This included the construction of quays and railways along the banks outside the Old Town. There was Packhus Quay and Skeppsbro Quay as well as the docks at Stora and Lilla Bommen which were extensions of Stora and Ostra Hamnkanalerna, (the Large and East Dock Canals). At Packhus Quay a new City customs house was built in 1864. At the turn of the century both quays were developed into a centre for trade and shipping with a number of expensive buildings, many of which were owned by the leading shipping companies. Skeppsbro quay with its wooden pier and stone pier later became primarily a mooring point for boats operating along the coasts.

East of the Old Town, Gullberg Quay was built in the 1880s. This was used primarily for barge traffic and canal boats.

Further west the old timber jetties and sheds were gradually replaced by stone quays and large timber warehouses. Stigberg Quay, which was opened in 1910, had a depth of 8 metres and it was the first one built that could take ocean-going vessels. This became the mooring point for the passenger ships of the Swedish American Line. In 1912 the so-called 'Concrete Shed' or 'American Shed' was built here, the first shed ever made of concrete.

Next to Stigberg Quay, Gothenburg's Fish Dock and auction hall, warehouses, etc. were built between 1908 and 1914. For a long time Gothenburg had been an important centre for fishing on the west coast. During the 1800s the selling of fish was primarily done at Rosenlund Quay and the fish hall inside the moat. After the construction of the new Fish Dock, west coast fishing became even more concentrated on Gothenburg and in the 1920s a canning factory and freezing plant were built next to the new Fish Dock.

The last big expansion along the south bank of the river was Majnabbe Dock built alongside the East India Company's old dock area. The first part of the quay and the terminal were built in 1951 for passenger vessels to England. The area was expanded in 1955 to accommodate the traffic to Denmark.

6   *The Alvsborg Bridge and the inner parts of the Port, built before 1950. (Port of Gothenburg AB, B W bild).*

North Bank of River

Since the middle of the 19th Century the north bank of the river has been used for industries run by the people of Gothenburg.

It was the shipyards and the mechanical workshops which dominated and which were in existence long before the docks. The first one, the Lindholmen Shipyard, was started in 1845. It was developed into a large shipyard in the 1870s and a working class residential area grew up around it. Further west a mechanical workshop was built in 1852, which later became the Eriksberg Shipyard. Opposite the Old Town a shipyard called Gotaverken (now called Cityvarvet) was built in 1867. It was originally on the south bank of the river in the Old Town and after moving it soon became the biggest in Gothenburg.

At the beginning of this century the shipyards, along with two large timber factories and two mills, occupied a large area of the river banks.

Expansion of the Port began in 1906 after annexation. The first one was Sannegard Dock, built in a natural cove between two hills, Sorhallsberget and Slottsberget. The dock was 7-8 metres deep and was used for the handling of imported coal and coke.

In 1922 a second dock was completed. It was a free port which was built as part of the Port Plan of 1912. The Free Port included a pier and two large stone warehouses. The area was expanded in various stages later on. Next to the Free Port, Lundby Dock was built in 1952 and this became part of the Free Port.

West of the present Cityvarvet, where one of the timber factories previously stood, Lindholm Dock was built in 1939. It had two docks, a shed and a silo.

7   *Sannegarden Dock and the hill Slottsbergget with residential buildings. To the right: parts of the former Lindholmen shipyard and the dry dock. (City Building Office, 1985).*

8   *The old town looking north west with the moat in the foreground
and the Large Dock Canal on the right flowing into the Gota river.
On the north bank opposite the old town and the Stone Pier is the
shipyard Cityvarvet, former Gotaverken. Compare with illustration
number 3, 1985. (City Building Office).*

9 *Stora Bommen Dock and Packhus Square with buildings erected for the leading shipping companies. 1987. (Historical Museum).*

10 *Lilla Bommen Dock and typical dock shed of timber construction. In the dock boats belonging to the ship museum. 1979. (Historical Museum).*

Development (1950-1985)

Modern methods of handling goods requiring large, rational areas and the import of oil has to a large extent controlled the development of the Port after 1955. For the south bank of the river this has meant cutbacks and co-ordination of shipping. On the north bank widespread expansion was taking place westwards.

During this period, communication between the banks has been improved by the building of the Alvsborg Bridge and the Tingstad Tunnel which were completed between 1966 and 1968. At the same time all the ferry routes across the river, except one, were discontinued.

On the south bank the harbour traffic has been closed down or has been concentrated in certain areas. The railway has been removed and along the former tracks now runs a large dual-carriageway—Gotaleden.

The only real regular traffic of any great significance is that of Stena Line which leases the majority of the quays west of the Old Town. However, the Fish Dock is still operating. On the quays in the Old Town there are some passenger boats which operate in the Archipelago and on the Gata canal during the summer.

After 1950 the expansion of the Port was concentrated on the westerly parts of the north bank. Three oil harbours were built. A small oil harbour dating from the 1930s, Rya Dock, has been expanded. Adjacent to this, Skarvik Dock was built betweeen 1951 and 1957 to handle the greatly increased oil traffic. Between 1962 and 1967 Tor Dock was built at the mouth of the river for the import of crude oil to the newly built refineries. Vessels of up to 200,000 tonnes could be accommodated at the 360 metre long pier.

Scandia Dock and Alvsborg Dock were built by extensive in-filling of the Alvsborg Fiord. Scandia Dock, built between 1962 and 1968, was primarily for container traffic. There are also terminals for passenger traffic to Germany and England. Alvsborg Dock, which was completed in 1975, was built originally as a ro/ro dock. Altogether these large docks have a total quay length of 2,600 metres.

The three large shipyards Gotaverken, Lindholmen and Eriksberg all increased their operations from the 1940s and required larger areas. Parts of the hills were blasted away to create the extra accommodation. A great deal of expansion was planned in the 1960s which would have involved demolishing parts of Slottberget and Sorhallsberget but this project was never carried out.

In 1963 Gotaverkken moved major sections of its business to Arendal outside Alvsborg Dock. Between 1975 and 1976 the closing down of the combined Eriksberg-Lindholmen Ship-yards was started.

Old Docks on South Bank

The oldest dock area, Stora Hamnkanalen in the centre of the Old Town, has been completely preserved in its original design. The quay reinforcements and a couple of iron bridges, however, date back to the middle of the 19th Century. Next to the old dock is the East India Company's large stone building erected between 1750 and 1762 and used as a museum since the 1860s.

The quays in the Old Town, running out towards the river, have mostly looked the same as they did when they were built around 1860 Today they are used primarily for tourists pleasure trips and museum ships.

A particularly valuable area is Packhus Quay and the adjoining Packhus Square which are situated where Stora Hamnkanalen runs into the river. Here there are a number of well preserved buildings which have connections with the port. At the quayside is the Customs and Packing House dating from 1864 (Architect, A. Heilborn) and a wooden shed dating from the turn of the century. Along the east side of Packhus Square with their frontage facing towards the docks, are: the former Brostrom Company's Building with its monumental facade in the classical style of the 1920s, (Architect, S. Steen), the Transatlantic Building erected in 1942, (Architect, S. Steen), and the Herza Building from 1901, (Architect, L. Enders). The Gothenburg Maritime College is built on the hillside (1859-1970) Between the quays and Packhus Square there is the busy Gotaleden road which has a distinct screening effect.

The dock at Lilla Bommen, east of Packhus Quay, is used as a guest harbour for small boats and for the sale of boats. There is also a museum with the museum ship, the barge 'Viking' plus many others. In addition, there is a ship museum which is housed in a dock wooden shed erected in the 1920s.

Lilla Bommen is the end of the Gothenburg promenade street Gota Square- Kungsport Avenue-Ostra Hamngatan down towards the river and has, therefore, featured a great deal in town planning in recent years. According to the plans the present operations at Lilla Bommen Dock will continue to be developed. On the east Quay considerable changes are envisaged. Currently, there is a large office complex being built for IBM and Skanska. Among other things there will be a 20-storey block built which will radically break the traditional, small-scale building development which is typical for the inner city of Gothenburg and the older dock areas.

Gullberg Quay in the east and its adjoining areas are at present undergoing the last stages of a comprehensive renewal. Here the timber warehouses of 1-2 storeys have been replaced by office blocks of 5-6 storeys. An older preserved building is the Tobacco Monopoly's stone building dating from the 1920s, (Architect, C Johansson), which is an important feature in the dock environment. How this will be used in the future is not yet clear.

The outer areas of Gullberg Quay have been given over to old sloops and other boats of varying descriptions. Between the quays and the office area are cycle paths, flower-beds, etc.

The quays to the west of the Old Town with the exception of a small area are leased by Stena Line for its ferry traffic to Denmark and Germany. The whole area from Masthugg Quay to Stigberg Quay is closed off and all the old buildings have been demolished. At Majnabbe, Stena Line uses the terminal which was erected for the ferry services to England in the 1950s.

Between Stena Line's two areas is the Fish Dock, which retains its genuine character, and the 'America Shed' which will be converted into offices and restaurants.

Out in the west, next to the Alvsborg Bridge, is a cultural reserve which was created at the East India Company's old dock. Here there is a timber warehouse from the 18th Century (converted into a restaurant), the Company's offices from the 19th Century and residential buildings from the 18th Century which together with the former Carnegie Brewery buildings from the 19th Century make this dock district a culturally and historically valuable conservation area.

11 *The Free Port, Lundby Dock and the shipyard Cityvarvet. In the background the old town with Lilla Bommen to the left. 1985. (Port of Gothenburg AB, B W bild).*

12 *Scandia Dock and Alvsborg Dock, the most modern parts of the Port. On the small islands outside Alvsborg Dock there is a fort (Nya Alvsborg), built in 17th Century for the defence of Gothenburg. (Port of Gothenburg, AB, B W bild, 1985)*

13 *Cityvarvet, (formerly the shipyard Gotave-ken) seen from the Stone Pier. 1987. (Historical Museum).*

14 *Machine shops at the Eriksberg Shipyard and the mill 'Tre Lejon'. View from Majnabbe Dock. (Historical Museum, 1987).*

Harbour along North Bank

From the south bank's quays and streets one has a fine view over the north bank's continuous stretches of docks, former shipyards and harbours. Green belt and residential areas stretch right down to the river at only one point. This is at Sannegard Dock where the hills, Slottberget and Sorhallsberget, form an effective break and a very characteristic contribution to the locality.

Cityvarvet (formerly Gotaverken) is still operated partly as a ship repair yard but part of the area is used for other business. The buildings form a dense industrial environment. The workshop dates from 1906 whilst the other buildings date mostly from the period 1930-1950. Renovation of the buildings for new uses is being planned.

At the Erikberg Shipyard business ceased in 1979. Parts of the dock buildings have been demolished but there is still a machine shop from 1922, a slipway from 1940, (the lower level of which contains workshops, staff areas and offices), a dry dock with a gantry crane dating from 1959 as well as piers and quays. With the extensive closure of the shipyards between 1975 and 1980, it was suggested that parts of Eriksberg should be turned into a dock museum. The idea has still not materialised but some of the buildings have been declared industrially and historically interesting and part of the area will possibly be preserved.

When the Lindholmen Shipyard was closed down, all the older dock buildings were demolished. The dock areas in the east are now used by 'Project Lindholmen' for training and industrial development. The west part with its dry dock and shipyard workers' houses has, however, been preserved. The dry dock, which is partly built into the hillside, was constructed in 1875 and was later expanded. The residential buildings, detached timber houses, were built between 1875 and 1900. Especially well-preserved are the buildings on the Slottsberget hill near the river. The houses and the dry docks have been declared as culturally and historically very valuable and for the last 15 years they have been part of the local authority preservation programme.

The Free Port, including Lundby Dock, is used today as a free port, mainly for banana imports. The first buildings from the 1920s have been demolished. Still standing is a well-preserved stone shed erected in 1944. On the oldest pier are railway tracks with the turntables still preserved. The Free Port is also one place in the Port where you can still see a large number of cranes in use. The inner north pier is currently being rebuilt as a ferry terminal for the rail ferry between Gothenburg and Freddrikshamn.

The two smaller docks, Lindholm and Sannegard, are also in use. From the Lindholm Dock with its two goods sheds, Volvo cars are shipped abroad while Sannegard Dock is used for the import of coal and salt.

When the Shipyards' activities declined in the 1970s there was a discussion as to how these areas should be used. At that time there was a surplus of housing accommodation in Gothenburg. However, in recent years Gothenburg has been experiencing a shortage of housing and the north bank has been discussed as a possible residential area. Since 1986 a comprehensive plan for the whole area between the Gota Alv Bridge and the Alvsborg Bridge is being considered. Proposals for building entire city areas with both housing accommodation and working environment are being debated. At the same time the possibility is also being discussed of preserving the most interesting industrial areas, including Eriksberg.

15 *Lindholmen-Slottsberget with shipyard workers' houses before renovation. (Historical Museum, 1972).*

Port Administration

During the period 1824 to 1876 the Swedish state and the local government together had the responsibility for the Port of Gothenburg. Between 1876 and 1985 the port was managed as a port of the local administration (Hamnstyrelsen). In 1985 a special company 'The Port of Gothenburg AB' was founded by the local authority. Shortly after that the pension funds SPP/AMF bought the real estates in the port and they are now rented by the company.

Acknowledgements

The original was written by architect Gudrun Lonnroth, of the Historical Museum of Gothenburg with the co-operation of Dr Thomas Thieme of the Maritime Museum of Gothenburg. The translation from Swedish was by Knowles & O'Malley AB.

The photographs and illustrations have been supplied by Gothenburg Historical Museum, Port of Gothenburg AB and City Building Office to whom we are deeply grateful for permission to reproduce.

Bibliography

Atterman, Artur: Goteborgs stadsfullmaktige 1850-1962 (I pp 347-352, II pp 275-80), Goteborg 1963.

Bensdorff, Leo: Goteborgs Hamn genom tiderna, Goteborg 1932.

Hansson, Sigurd: Goteborg som storhamn, Svenska Turistforeningens arsskrift (pp 42-66), 1932.

Petterson, Knut: Goteborgs hamn, Goteborg en overskift vid 300-arsjubileet 1923.

Petterson, Knut: Goteborgs hamn under 50-arsperioden 1882-1932, Tekniska samfundets i Goteborg minnesskrift (pp 309-348), 1932.

Paulson, Lars: Redogorelse for vissa utbyggnader i Goteborgs hamn under aren 1962-1986 (compendium with technical details, out of print).

Periodical publications

Hamn (News from the Port of Gothenburg)

Svenska Sjofartstidning (Swedish Shipping Gazette)

Annual reports from the Port of Gothenburg AB

Archives

The Historical Museum of Gotenburg (Goteborgs Historiska Museum)

The Maritime Museum of Gotenburg (sjofartsmusseet i Goteborg)

The Port of Gothenburg AB

16 *View from Lilla Bommen and the barge 'Viking', towards the Free Port. 1987. (Historical Museum).*

West Germany

Port of Duisburg-Ruhrort

Mr Axel Fohl

Introductory Note

If one tries to give an outline of the history, development and future of the Port of Duisburg-Ruhrort, one does so at an especially crucial period in its more than 250 year life cycle. The handling of bulk products, mainly those of the extractive industries such as the export of coal from the Rhine-Ruhr coal mining district and the import of iron ore of European and overseas origin is rapidly and drastically coming to an end. Losing this raison d'etre, the port today has to decide on new policies and this, of course, can be a crucial era in the life of the majority of physical remains from the history of the harbour system. Concerning this heritage as far as it can be defined within the framework of industrial archaeology, we can, at the moment, only point out that recording and listing activities have just begun. Industrial archaeology as such, together with governmental listing and protecting procedures have had a much later start in the Federal Republic of Germany than for example in Great Britain and consequently the treasure of knowledge—as mirrored in a wealth of British publications such as the splendid inventory of the London docklands—is much less and incomplete.

With regard to industrial monuments within the port listing and legislative protective measures have only been taken sparsely, as for example with some of the grain mills in the port. But before we can discuss the historic features, an outline is given below of the general history of the area.

The Harbour Prior to 1905

To describe today what the half million inhabitants of Duisburg call the largest river port system in the world, one has to bear in mind that this system is composed of a nucleus of publicly owned harbour components surrounded by a wealth of industry-owned private ports with names like Thyssen, Krupp and Mannesmann amongst them, as well as that of the now unified Ruhrkohle AG, an amalgamation of West German coal mining enterprises. The emphasis of this contribution will be on the publicly owned part of the whole system, simply because up to the turn of the century public activities have formed the main impetus in the development of the harbour. Here a distinction has to be made between the history of the earlier and more important part of the town of Ruhrort and its southern counterpart, that of the town of Duisburg. Although both communities were equally favourably situated on the

western borderline of what was later to become the heavy-industry-area of the Ruhrgebiet, the town of Ruhrort had the advantage of being positioned on the banks of both the rivers Ruhr and Rhine.

In the course of river changes Duisburg has over the centuries lost its position immediately alongside the river Rhine and in consequence it was the town of Ruhrort, which translated means the place where the river Ruhr flows into the Rhine, that was able to use its topographical advantages. Therefore the political territorial background gave Ruhrort better chances. Both the territory of the Duchy of Cleve and the Country of Mark had come into the possession of Brandenburg-Prussia, thus bringing the era of the coalfields and the harbour site under one reign. At the beginning of the 18th century under the reign of Frederick William I of Prussia (1713-1740), a small harbour basin of about 10 metres wide was constructed on the right bank of the River Ruhr just before it flows into the Rhine, west of the town of Ruhrort. The 1753 plan of the harbour shows its gradual enlargement as well as the long rectangular spaces parallel to the river where coal was stored. This was brought down the Ruhr in small wooden barges called Aaks and reloaded for up-and-downstream transport in the larger vessels on the River Rhine. Traffic increased considerably, from 1776 when Prussia built locks alongside the weirs in the river Ruhr, thus facilitating coal transport from the mines up river down to Ruhrort harbour.

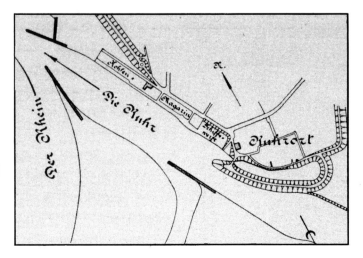

1   *Ruhrort Harbour in the 18th century. The 1753 map shows the early basin on the bank of the river Ruhr just before it flows into the Rhine west of the town of Ruhrort. The rectangular spaces were used for the storage of coal brought in small wooden barges called 'Aaks'. (Hafag)*

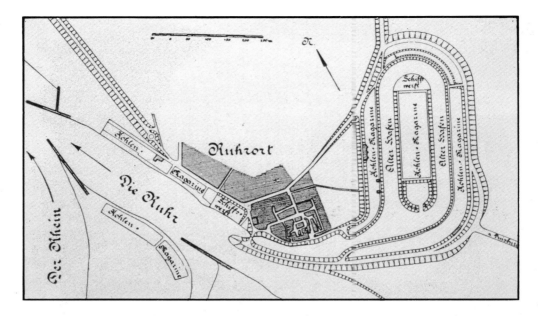

2 *Duisburg-Ruhrort Harbour early l9th century. The first dock (Inselhafen) independent of the river was built between l820 and l825, forming an elliptical shaped basin of about 1500m long with 7 hectares of water area. (Hafag).*

The first real dock independent of the river, (because of state ownership it was called fiscal harbour) was created between 1820 and 1825, forming an elliptical basin of about 1500 metres in length with around seven hectares of water surface. This enlargement was mainly financed out of the Ruhrschiffahrtsfond made up of contributions from the barge owners using the Ruhr river. The new harbour Inselhafen considerably increased coal traffic; by 1834 it had risen from 160000 to 340000 tonnes per year. The same Ruhrort coal merchants who had pressed for the 1820 extension now obtained permission from Prussia for a second expansion that brought the water surface of the Ruhrort harbour up to a total of 117 hectares in 1842. The new, so-called lock harbour with a length of one kilometer and a bottom width of 25-30 metres followed the curve of the elliptical 1820 port basin and was linked to it by a channel east of the town of Ruhrort.

In 1848 the Cologne-Minden railway was extended to Ruhrort harbour. By 1853, 480703 tonnes of coal were transported to Ruhrort, roughly a quarter of it by railway, the rest by boat. The large scale use of the steam engine from about 1850 led to a sharp increase in the hoisting capacity in the Ruhr coal mines. The steel industry started to develop on the coalfields, producing iron and steel that had to be shipped and needed iron ore and other bulk goods to be imported. As a consequence, Ruhrort harbour had to be enlarged again. North and South harbour extensions increased the length of the port to a total of 4.5 kilometres and its surface area of c. 30 hectares. Both of these docks were built during 1860 to 1868 and they had the extraordinary bottom width of about 70 metres thus reaching a size nearly big enough for todays demands. The river Ruhr was diverted to the south and separated from the harbour mouth by a jetty. By 1869 more than 1.3 million tonnes of coal were brought to Ruhrort to be shipped from there. The industrial boom that followed the Franco-German war led to the construction of the docks Kaiserhafen in 1872-1890 southeast of North and South docks. Again the river Ruhr was diverted to the south to create a separate entry to the new part of the Ruhrort harbour so that there were now three outlets of the Ruhr and port system towards the Rhine.

The port had increased its size to 7.5 kilometers in length and 51.3 hectares in surface area. More than 60 kilometers of railway lines signified the end of the river transport of coal down the Ruhr, that virtually ended in 1890 when more than 2.6 million tonnes of coal were brought into Ruhrort by rail. The same year also meant the peak of the harbour development in Ruhrort. The town that by now had nearly 40,000 inhabitants had developed to the boundaries of its territorial and administrative capacity. In 1905, when it united with the towns of Duisburg and Meiderich, where parts of its latest harbour extensions were situated, Ruhrort had huge steel works, conveying more than 5 million tonnes of coal to southern Germany, the Netherlands, Belgium and to the North Sea and received more than 1.3 million tonnes of iron ore from Spain, Sweden, Italy and Canada. It was in these peak years of the development of the heavy industry of the Deutsches Reich that three large new harbour basins were created to the east and parallel to the three existing basins on Ruhrort territory. These three new docks A, B and C built from 1903 to 1908 were—with few exceptions—exclusively planned for coal transhipment. Again the river Ruhr was pushed south, basins A, B and C got their own direct outlet to the River Rhine, the so-called harbour channel and a new railway terminal at their northern end. Electrical dump-tyre-coal-barrows were installed which raised the annual capacity up to 18.7 million tonnes of coal in 1913.

Since 1905, Ruhrort, Meiderich and Duisburg have formed a new municipal unit. Duisburg had brought into the marriage its own, much smaller harbour system, generated in competition with the port of Ruhrort. Changes in the flow of the river Rhine had forced the prosperous medieval trade centre of Duisburg to dig a canal of two kilometers length and 8.5 metres width, which was completed in 1832 linking town and river once more. Since this canal had no connection with the coal carrying river Ruhr, a second canal was built in 1840-1844 by the Rhein-Ruhr-Kanal-Aktien-Verein. Both channels were enlarged and combined with railway lines in later years. After the connection to the river Ruhr was abandoned again, the canal to the Rhine was made into a harbour basin, the Aussenhafen Outer harbour). In this way the town of Duisburg had two basins at right angles to the river Rhine, immediately fronting the town centre. During the period 1882-1893, the whole system was enlarged to 34.7 hectares. During 1895-1898, it was supplemented by a third independent harbour basin to the north, the Parallelhafen that was to handle coal and ore in competition to Ruhrort.

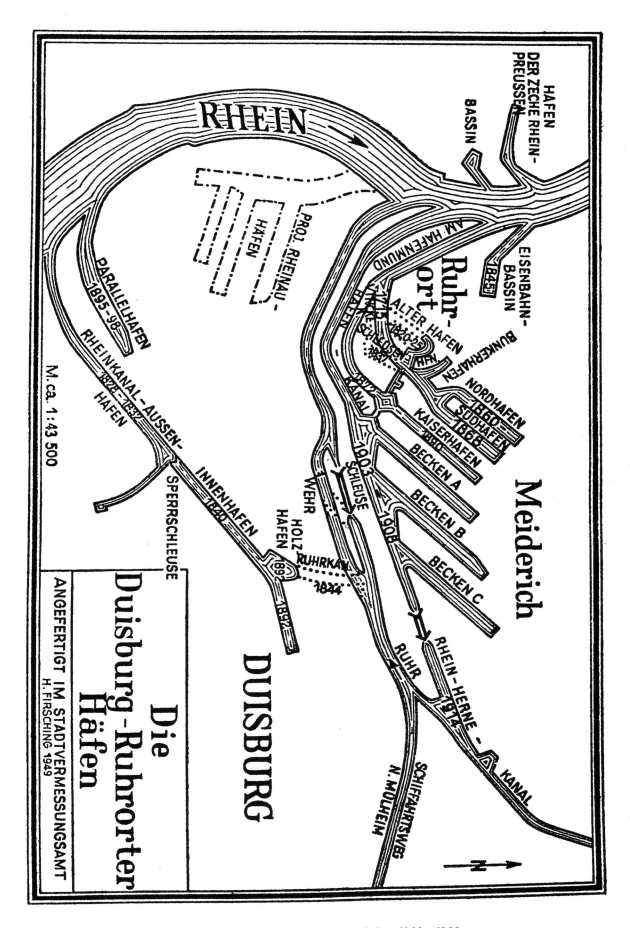

3 *Three new Docks, A, B and C were built from 1903 to 1908.*
primarily for coal transhipment. The docks had their own outlet
channel (Rhein-Herne-Kanal) to the Rhine. Today the sites are
referred to as the 'Petrol Island' for handling imported crude oil and
the 'Coal Island' with a huge coal mining and loading plant. (Hafag).

Development Since 1905

The unification of 1905 stopped Duisburg's plans for three large basins to the north of the town. In 1908 the newly unified town acquired a third harbour system, the Hochfelder Hafen with three basins from the Rheinische Bahngesellschaft, a Railway company, and in the same year the Railway Harbour north of Ruhrort, built for the ferry linking both banks of the river Rhine. Since 1912 all four port systems have come together under the common roof of the Verwaltung der Duisburg-Ruhrorter Hafen (Administration of the Harbours), founded in 1905 to end the economically dangerous and useless competitive development that could only lead to surplus capacities. For a short time (up to 1914) the Duisburg-Ruhrort harbour held the peak of its capacity in terms of water areas. Even after a part of the elliptical Old Harbour and the first enlargement, the Lock Harbour had been partially filled in, the public parts of the system comprised 189 hectares of water space 477.4 hectares of harbour-owned land, 43.8 kilometers of docks and 311 kilometers of railway lines.

From the start the administration as a joint venture of the Prussian state and the town of Duisburg, remained owner of all land, working with a system of long term leases with its individual partners. From 1926 the harbour administration was converted into a privately organized limited company. Since 1960, the Federal Republic of Germany, the state of North-rhine-Westphalia and the town of Duisburg are each one third shareholders of the capital of 30 million Deutsch-marks.

The First World War had brought all expansion activities to a halt and the period from 1918 to the end of the Ruhr occupation saw a reduction of trading which lowered the amount of cargo handled to 3.4 million tonnes in 1923. After the London Treaty the figures nearly reached their pre-war status again with 22.5 million tonnes in 1925. Just before the war, in 1914, the Duisburg harbour system had—by means of the eastward running Rhein-Herne-Kanal—obtained its link with the rapidly growing German inland canal system. This canal today enables the so-called Euroship of 1300 tonnes to reach every German inland waterway or port north, south or east. Nevertheless not more than 5-10% of traffic is handled with canal routes today, whereas 60-80% go to or come from Rotterdam, between 10 and 20% to or from other estuary ports

and 5 to 15% to southern ports on the river Rhine. Nowadays far more goods are coming in than are going out since the export of coal reduced enormously. Imports, such as mineral oil, imported coal, fertilizer or ore currently dominate although the latter, due to problems in the steel industry of the Ruhrgebiet, is reduced too. On the other hand, a good part of the ore import has always been managed by the rather large number of privately owned harbours south and north of Duisburg.

For the publicly owned port system the following table, giving the figure for the development of transport of different goods (in tonnes) between 1936 and 1983, distinctly shows the decrease in coal traffic:

	1936	1978	1983
Ore	1,092016	10.108939	6,250199
Mineral Oil	208617	4,400537	3,500773
Coal	12,091988	5,012292	3,949035
Iron/Steel/ Non-ferrous metals	407844	3,082038	2,665168
Stone/Sand/Clay/etc	2,036130	1,505629	1,334716
Other	1,396028	1,235800	1,492889
Local traffic	133608	418609	291137
Total	17,366231	25,763844	19,483917

The part of the port that originally handled coal started looking for new cargoes after World War II. The Kaiserhafen dock was mostly filled in to create space between the Sudhafen and basin A. Today this site is called Petrol Island and handles imported crude oil (since 1954) which is transported to the Gelsenkirchen refineries. Since 1959 crude oil has been pumped via a pipeline from Wilhelmshaven in the north. The site between basins A and B, further east, remains an area devoted to coal. On this so-called Coal Island a huge coal mixing and loading plant has been erected by Ruhrkohle.

4　*Duisburg-Ruhrort Harbour 1890. The 1842 Lock Harbour linked to the first dock by a channel east of the town of Ruhrort. The river Ruhr was diverted to the south and separated from the harbour mouth by a jetty. (Hafag)*

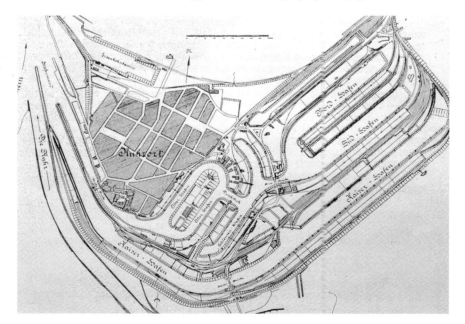

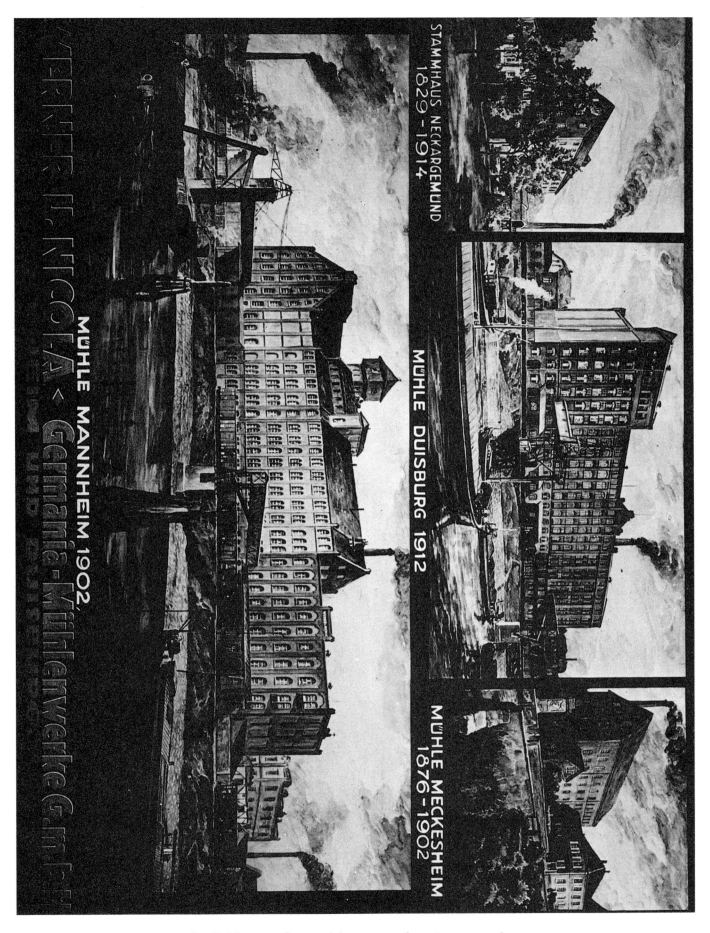

5 *Duisburg was Germany's largest centre for grain storage and corn milling at the turn of this century. The attractive seven storey Germania Mill of 1912 exemplifies the migration of the grain milling business from rural areas towards the river ports. (Hafag).*

6 *Haniel Wharf warehouse of 1862 is one of two surviving buildings of steam transport on the northern part of the elliptical basin (Inselhafen) of 1820-25. In 1829 Haniel built his first steam boat and opened a new era of transport history on the river Rhine. (Haniel Museum).*

Industrial Archaeology

Since the end of the seventies, steel followed coal into a crisis and from that time there dates a new policy for the revitalisation of Duisburg-Ruhrort harbour. It aims more and more for high-quality iron and steel goods from all over the Ruhrgebiet, such as high-quality steel coils, requiring roofed over loading areas. Container and roll-on/roll-off trailer traffic is another new enterprise that has led to changes in the appearance of both docks and land spaces. The quay walls were mainly affected by two unrelated developments River improvements speeded up the flow of the River Rhine thus creating an erosion problem that led to a considerable deepening of the river bed. In relation to the river, the Duisburg-Ruhrort port system level rose higher and higher, adding to already substantial problems with water levels. The water level in Duisburg can rise or fall by more than 8 metres. These changes caused the bottom of the port basins to be deepened again and again, thus undermining the original basin walls. For a long time it was thought that another problem—the continuous sinking of the harbour area, because of mining activities underneath the port, could be advantageously used to compensate for the sinking of the river bed. Since the 1950s attempts have been made to guide mining activities underneath Duisburg-Ruhrort harbour in such a way that an adjustment to the river level was achieved. Geological irregularities however prevented that scheme from succeeding and consequently the dock walls have to be re-constructed where and if need arises. A large part of the walls have thus been renewed and further renovations are under way. Another future measure will include the filling in of some of the older basins to gain urgently needed space, since in the Duisburg-Ruhrort harbour the relation between land and water spaces is very much to the disadvantage of land areas owing to the original character of the port as a bulk goods handling harbour with very fast turnover. To bring in new enterprises, the administration feels that it has to offer large and unimpeded spaces for every imaginable purpose.

As initially stated, all we can give at the moment, is a rather fragmentary report on historic features of archaeological interest. What strikes a visitor first, when he turns toward the harbour from the Duisburg town centre, is a line of multi-storey brick buildings, whose roofs rise up above the silhouette of the neo-renaissance town hall and the Gothic St Salvator church. These are grain stores and grain mills, mostly dating from around 1900. Since 1860 mills started to gather round the Innenhafen of Duisburg, making the town West Germany's largest centre for grain storage and corn milling at the turn of the century. Near the east end of the Innenhafen the seven storey, 32-bay Germania Mill of 1912 exemplifies the migration of the grain milling business from rural areas towards the big river ports. The multiple advertising signs show the firm's two mills in the harbour of Duisburg and Mannheim as well as its earlier, much smaller enterprises deep in southern Germany. In 1900 Duisburg harbour handled more than 4.75 million tonnes of goods, 0.6 million of it was grain. Next to the Germania Mill of Werner and Nicola, the huge West-German combine of Wehrhahn placed a six storey, 14-bay mill, which differs in its appearance from that of its neighbour, although it is nearly contemporary. The influence of what in Germany at the time was called Neues Bauen—(New Building)—lead to a dispensation with historical motives in the building's external appearance, that became a standard style in industrial buildings before World War I. This is clearly shown by another group of grain stores and mills that follow to the west along the Innenhafen dock of Duisburg.

Next to grain, timber was handled in Duisburg harbour in great quantities, a good part of which went down the mines of the Ruhrgebiet before steel was introduced. This activity took place around the Holzhafen (Wood Harbour)—of 1892 but has left no traces in the shape of building structures. From the beginning, Duisburg was linked to Ruhrort in the north-west

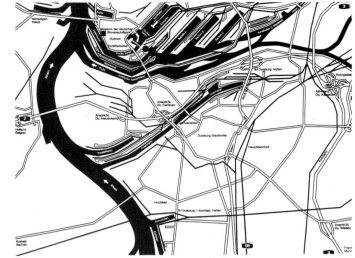

7   *Schwanentor (Swans Gate) Bridge was opened in 1950 to replace the old bascule bridge of 1903. It is one of only two lifting bridges remaining in the port. (Hafag).*

8   *Duisburg-Ruhrort Port in the 1980s. General map of the growing harbour showing the oldest remains of the port system in Ruhrort and Hochfeld on the western side of Duisburg, the region of the port's chemical and steel industries. (Hafag).*

by bridges, first across the Ruhr and later also across the growing harbour. Originally all of them had been movable bridges of different types but later road construction dispensed with most of them except two lift bridges. Schwanentor (Swan's Gate) bridge was opened in November 1950 to replace the old bascule bridge of 1903. A second one Marientor-brucke—(Mary's Gate) Bridge comprising parts from the 1920s and 1930s, the moving part being renewed in the 1970s. The western part of Duisburg harbour is sector of town called Hochfeld (High Field) south of the Ausenhafen, was originally characterised by the steelworks of the second half of last century that profited both from the vicinity of coal and transport facilities. Today Demag (part of the Mannesmann combine) produces machinery there. Next to it the firm of Matthes and Weber (part of the Henkel chemical empire) represents the tradition of the Duisburg chemical industries, together with others which started business in 1832.

Crossing Duisburg's Innen and Ausenhafen, the river Ruhr, the Hafenkanal, Kaiserhafen and Vinckekanal (named after the Prussian official, who greatly furthered the aims of the Ruhrort merchants in the early 19th century), we come to the oldest remains of the Duisburg-Ruhrort port system in Ruhrort. Like a question mark the remaining northern part of the elliptical Inselhafen of 1820-25 curves to the right. It is one of the few places where one can see the original embankment walls sloping down into the now quiet waters of this part of the harbour. Here the history of the firm of Franz Haniel mingles visibly with that of the port itself, thus mirroring a lifelong connection between firm and town history. As one of the very early firms Haniel developed out of small scale coal trading into a mining, iron and steel, machine and shipbuilding and transport business. In 1817 Franz Haniel had helped James Watt junior with engine trouble in his steamer Caledonia on the River Rhine. Sensing the importance of steam transport, Haniel in 1829 built his first steam boat and opened a new era of transport history on the river Rhine. As a physical com-memoration to this development, two brick buildings have survived north of the Inselhafen dating respectively from 1862 and 1871. The 1862 building of two and a half storeys, served as a warehouse for the Haniel Wharf, situated on the island in the middle of the 1820 harbour basin. In 1871 the firm of Haniel built a new forge at right angles to the circular dock.

Of the several generations of coal-dumping devices only a few have survived due to the restructuring and the leasing policy of the Hafag administration. Since in most parts of the harbour there are no individual owners of the land everything is cleared away when the lessee moves out. Of the 90 cranes and loading bridges, up to 40 tonnes, only a few date back before World War II. There have never been hydraulic systems in the port. The bridges were worked individually. For a time, electricity was generated locally, but for more than 50 years now it has been provided by Rheinisch-Westfalische Elektrizitatswerke, the large West German power supplier. To the east the port is separated from the levels of the river Ruhr and the Rhein-Herne-Kanal by two locks, both post World War II structures with some components from earlier dates.

This article has been more of a provisional profile rather than a resume of years of survey. It is to be hoped however, that an inventory will be compiled during the next two years. Perhaps the information given here will be sufficient to create an interest in this fascinating harbour landscape that is generally called Germany's most western seaport.

9   *In 1871 the firm of Haniel built a forge at right angles to the elliptical dock with similar architecture to that of the warehouse building. (Haniel Museum).*

Further Information

There are two museums concerned mainly with the Duisburg-Ruhrort harbour history. The first is Museum der Deutschen Binnenschiffahrt (Museum of German Inland Navigation) with a beautiful large scale model of the whole harbour. The museum centres its exhibitions around the development of ships and ship building on the rivers. Since 1974, it owns, as an annexe, a 1922 paddle steamer called Oscar Huber that serves as a floating museum of tug service history.

A company owned museum is that of Franz Haniel and Cie GmbH, whose documents and exhibits refer in many ways to the development of Duisburg-Ruhrort ports. The building that houses the second museum is one of the 18th century warehouses of the Ruhrort merchant dynasty of the Haniels.

Further information can be obtained at the offices of Hafag, the administrative organisation of the public harbour system. A wealth of material is to be found in the communal archives of Duisburg as well as at the Hafag and at Haniel and Cie GmbH, although permission to view has to be obtained from all of them.

The addresses are:

1   Museum der Deutschen Binnenschiffahrt
    Dammstrabe 11
    D-4100 Duisburg 13
    (Ruhrort)

2   Haniel Museum
    Haniel Platz 3
    D-4100 Duisburg 13
    (Ruhrort)

3   Duisburg-Ruhrorter Hafen AG (Hafag)
    Alte Ruhrorter Strasse 41-52
    D-4100 Duisburg 13
    (Ruhrort)

4   Stadtarchiv Duisburg
    Rathaus (Town Hall)
    41 Duisburg 1

Acknowledgment

We are deeply grateful to Museum der Deutschen Binnenschiffahrt, Haniel Museum, Duisburg-Ruhrorter Hafen AG (Hafag) and Stadtarchiv Duisburg, for their kind permission to reproduce photographs and illustrations.

Bibliography

1   Ernst Ottmann, Die Duisburg-Ruhrorter Hafen. Denkschrift zur Vollendung der in den Jahren 1903-1908 ausgefuhrten Hafen-Erweiterungen, Duisburg 1909.

2   Rhein-Ruhr-Hafenbetriebsverein e.V. Duisburg (ed.), Die Rhein-Ruhr-Hafen. Ein Fuhrer, Duisburg 1926.

3   Fritz Wilhelm Achilles, Rhein-Ruhr Hafen Duisburg, Duisburg 1985.

10   *The Museum of German Inland Navigation in Ruhrort centres its exhibitions around the development of ships and ship building. It owns the 1922 paddle steamer Oscar Huber that serves as a floating museum of tug service. (Museum der Deutschen Binnenschiffahrt).*

Index

Specific references within captions to illustrations and maps are included among the general textual references.

Acknowledgements

I wish to thank my institution, The Polytechnic of East London (formerly North East London Polytechnic) for its support and the Commission of the European Communities (EEC) for a small grant to initiate a study on the history and development of European Ports, in association with Professor Roland Baetens of the University of Antwerp in Belgium. His help and co-operation are most appreciated

I am deeply indebted to the following team of international experts for their assistance and contributions in preparing the book:

Mr Albert Himler
Technical Inspector
Mr G Thues
Chief Engineer/Director
Stad Antwerpen,
Belgium.

Dr Jorgen Sestoft
Department of Architecture
Royal Academy of Fine Arts
Copenhagen
Denmark

Mr Alain Demangeon
Architect
Paris
France

Mr Hans Meyer
Engineer of Urban Renewal
Port of Rotterdam
Holland

Mrs Gudrun Lonnroth
Architect
Goteborgs Historiska Museum
Goteborg
Sweden

Mr Axel Fohl
Monument Survey Officer
Rheinisches Amt Fur Denkmalpflege
Pulheim
West Germany

I am grateful for the supply of information and photographs by the Port Authorities and City Councils of: Antwerp, Rotterdam, Duisburg, Copenhagen and Gothenbourg. Their co-operations are most generous.

Special thanks are due to Ted Weedon for his enthusiastic assistance and proof reading, to Walter Evans for reading the initial articles, to Simon Pattle for his kind co-operation and assistance with the preparation of photographs and illustrations, to Margaret Youngman and Carole Anthony for their excellent typing of the manuscript, to Plilip Plumb for his general advice, to Barry Nottage for library assistance, to Nimmi Patel of ILEA for typesetting and to John Jones for artwork, to John Noble for preparing the index, and to my wife Irene Naib for her assistance and forbearance.

Assistance was also gratefully received from other European colleagues, including Prof. Franziska Bollery and Prof. Robert Verheigen of Holland, Prof. Frederic Bedoire and Prof. Hans Bjur of Sweden, Prof. John Michel of France and Dr Poul Stromstad of Denmark.

Dr S K Al Naib,
Head of Department of Civil Engineering
The Polytechnic of East London
Longbridge Road, Dagenham
Essex, RM8 2AS
England